Sectoral Innovation Systems in Africa

Sectoral Innovation Systems in Africa

Editors

Mammo Muchie and Angathevar Baskaran

AFRICA WORLD PRESS

TRENTON | LONDON | CAPE TOWN | NAIROBI | ADDIS ABABA | ASMARA | IBADAN | NEW DELHI

AFRICA WORLD PRESS
541 West Ingham Avenue | Suite B
Trenton, New Jersey 08638

Book design: Dawid Kahts
Cover art: Ananthy Baskaran

Library of Congress Cataloging-in-Publication Data

Names: Muchie, Mammo, editor. | Baskaran, Angathevar, editor.
Title: Sectoral innovation systems in Africa / editors, Mammo Muchie and Angathevar Baskaran.
Description: Trenton : Africa World Press, 2017. | Includes bibliographical references and index.
Identifiers: LCCN 2017008505 | ISBN 9781569025659 (hb : alk. paper) | ISBN 9781569025666 (pb : alk. paper)
Subjects: LCSH: Economic development--Effect of technological innovations on--Africa. | Sustainable development--Africa. | Appropriate technology--Africa.
Classification: LCC HC800.Z9 T468 2017 | DDC 338.064096--dc23
LC record available at https://lccn.loc.gov/2017008505

This book is dedicated to our parents

Wiezero Dessalu Asmetche and Muchie Getahun

Suseela Gandhimathi and Mayathevar Angathevar

and
All mothers who did sacrifice so much to raise children in Africa

Contents

Preface

Sectoral System of Innovation for Sustainable Development

Mammo Muchie & Angathevar Baskaran

There have been four distinct approaches to economic growth and development. The first is historical development that is best articulated in the history of development studies; the second is the Listian political economy approach that inspired import substitution industrialization and the developmental state. List introduced the political context explicitly and forthrightly to the challenges of economic development. The third is the Schumpeterian innovation approach to economic development. The latter introduced innovation as emerging with the dynamics of creative destruction by introducing explicitly sociological constraints to the dynamics of economic development. The last is the uneasy treatment of economic development by neo-classical mainstream economics. Neo-classical economics did not appreciate the emergence of development economics as a separate field and later in the 80s some neo-classical economists denounced the field with books such as the poverty of development economics (Deepak Lal) prompting reactions such as the Dilemmas of Economic Development by John Toye and others.

What this demonstrates is that there are a number of ways development economics is going through. Development studies has now become 70 years old. What has now emerged is the link of innovation system at various levels with development. We have now the innovation and development approach that is drawing lessons from both development studies and innovation systems. We think it will be useful to explore how the system of innovation has evolved, used and applied in order to filter through the epistemic virtue from it that can be employed to shed light on how currently it is possible to address through the SDGs how sustainable development can evolve into a new socially, economically and environmentally integrated development and innovation system.

A new unified approach that combines innovation system at various local, sectoral, national, regional and global levels with development also from the global to the national , regional and local areas is critically needed. There is a quest for new research now more than at any time in order to address the persistent poverty, unemplyment, inequality and climate change challenges confronting the world in general and Africa in particular. The contributions in this book address this important issue by taking mainly the innovation dynamics at a sectoral level without excluding the regional and national systems of innovation. We welcome all to read, engage and develop deeply the research knowledge frontier by promoting the new field of transformative innovation for sustainble development and human wellbeing.

Acknowledgements

We would like to thank the DST/NRF SARChI (Innovation and Development) at the Tshwane University of Technology, Pretoria, South Africa for initiating the Science, Technology, Innovation Studies series of books in collaboration with African World Press.

We recognize and appreciate the support given by Dr. Jideofor Adibe, Adonis & Abbey Publishers Ltd, London, United Kingdom, which enabled publication of this book.

We also would like to thank the African Journal of Science, Technology, Innovation and Development (AJSTID) and extend our fullest appreciation and thanks to all the contributing authors.

Our sincere thanks to Ananthy Baskaran, the cover designer, and Jan B. Mwesigwa (www.janmwesigwa.com) our initial layout artist for their hard work and dedication to the book project.

Finally, we express our deepest appreciation and gratitude to Mr. Kassahun Checole and Mr. Girma A. Demeke, Africa World Press, Inc. & The Red Sea Press, Inc. for their cooperation and support during the publication process. We also thank all the publications staff at Africa World Press, Inc. & The Red Sea Press, Inc.

The DST/NRF Research Chair at Tshwane University of Technology and Africa World Press will promote the science, technology, innovation studies book series applying the principle to promote innovation and knowledge on and for Africa.

Chapter 1

Introduction

Building Sectoral Level Competence & Innovation Systems in Africa

Mammo Muchie & Angathevar Baskaran

There are a variety of ways of defining innovation. The established definition takes two dimensions. The first is degree of novelty proving the innovation to be new to the producers such as the firms and industry, and the users and the market. The validation of the newness is identified by demonstrating usefulness by achieving the distinction and recognition of validation only by commercial economic gain. The type of innovation can be either a variety of products or process innovation.

There have been additional definitions that have broadened the economic bottom line for defining innovation by adding equal recognition of the social bottom line and the environmental bottom line.

Given we have now the Sustainable Development Goals and the global community needs to make a paradigm shift in going beyond merely economic growth measured by income choice to achieve sustainable development combining economical, social and environmental gains that can engage and re-center innovation to promote and realise sustainable development by achieving fully a combined measurable and clear outcome of economic, social and environmental output.

In Africa the key sector that requires transformation is agriculture. This means transforming agricultural sector will not and cannot follow the linear path as making the process organic is needed by re-thinking and re-designing a new validation criterion that is not purely economic and commercial. There is a need to craft a new incentive mechanism to promote the innovation validation to combine economics with the social and environmental gains. This re-defines innovation to accommodate the cost of innovation to the social and environment to be equally included upgrading the previous and the current dominant commercial economic gain as it has been argued by all traditional economists from Adam Smith to the present economic Nobel laureates. If the validation turns to be composite and not only commercially singular, a new social and environmental sensitive economics will be on the horizon.This has impact on all the definitions of innovation and to the varied actors that engage in interactive learning to forge the national innovation system from the local , sectoral , regional, national and global levels by arguing for a multi-level complementary incentive structures that are needed to be developed to motivate innovation to evolve in this synergised trajectory especially in the developing countries.

All along the issue has been with different shools of thought in economics. Issues that seem to preoccupy much of the economists hoping to create an alternative to the main-stream neo-classical economic framework appear to be understanding economic growth; short term and long term economic firm level and/or national performance, micro and meso level competitiveness, firm and national level productivity, economic catching up, leapfrogging, learning and knowledge creation and absorption in a given economic structure, and inter linkages between firm competitiveness and national competitiveness and productivity, symmetry and system building such as national, sectoral and other types of innovation systems. Since innovation is characterized by the process of creating and destroying, some economists including Veblen Thorstein (1857-1929) earlier on have not been open to the notion of innovation systems and symmetry. They focus more on asymmetry and system breaks than makes, associating innovation more or less with a dynamic that disrupts systems and symmetry rather than the opposite. The skepticism on innovation systems is understandable given that the available coordinating mechanisms such as the market, the state, the firm and others tend to operate in a way that may not facilitate symmetry and systems. In the real world, nearly all the elements, market, state, space, nation, sector, region, global, firm, technology, innovation and system do not exist without some interference from some variety of interaction. There is more anti-system than system in the context where making and forging development is on the agenda. The epistemic preference of system is more to convert anti-systemic and non-systemic phenomena into systems. Innovation is also disruptive. It is creative destruction, something new to the market, the firm, the user and producer always challenges what was old and what pre-exists already. Innovation is not constructive. A new validation criteria is needed to make innovation from being disruptive and destructive to be also constructive and inclusive.

The challenge is to create the sustainable and social-well-being anchored development economics of innovation systems from a context that the system and innovation are more likely to create disruptive construction that facilitates inclusive development in making simulatenous gains socially, environmentally and economically and not only economic gain at the an unacceptable cost of social and environemental loss. Thus the use of the integrating innovtive development system perspective is important as a focusing device to conceptualize, identify and select from the range of emerging forms of possible interactions, variations that are either emergent and to be made yet or already made, efficient or inefficient, strong or weak, necessary or contingent for generating outcomes and impacts on national economic development, productivity, competitiveness and overall better long term economic performance that never fails in including sustainably social wellbeing and environmental gains. In other words different innovation systems can be correlated with different outcomes and impacts on performance, productivity, competitiveness, capability, learning and competence or any combination of them. And how systems are constituted and the taxonomy and complexity of interactions, and the work to understand and explain them remains significant. To be sure, the real economic processes may deviate from what may be desirable, and from the way systems of innovation are forged. That does not invalidate the choice of how innovation systems emerge and are formed by the interaction of the structures, institutions, policies, knowledge and incentives in given environments and situations. Regardless of whether system building or not occurs in real economic systems, the local, sectoral, regional and national systems of innovation perspectives have been popularized.

They have constituted perhaps a significant development of appreciative theorizing. The main inquiry of systems of innovation at any level has focussed to understand the variations or differences in the innovation performance of countries that enter into explaining the long-term economic performance, productivity measured in such macro-economic variables as GDP and national competitiveness. Much of the work has been focused on industrialized economies not developing economies. Even in the developed economics innovation system research, the degree to which micro-level firm innovative capability, performance and competiveness can be aggregated to contribute to national innovative productivity, performance and competition has been analytically contentious. Appreciative theory in this innovation system genre has produced such terms as the knowledge-economy framework, the learning economy framework, and with the Globelics initiative, a further development has occurred. Globelics has combined together knowledge, innovation, learning and capability building and suggested research applicable to the problems of development and underdevelopment by translating innovation systems into : ' knowledge, learning, innovation and capacity, capability and competence building systems.' This opens up a possible line of inquiry where an alternative economic framework of combining " knowledge, learning, innovation and competence building' into an 'innovation and sustainable development systems' by including simulateneously the three inputs, namely the social, economic and environmental by addressing the problems and challenges of transition from underdevelopment to development for the developing world, the BRICS and others. If we proceed with the search and selection of an alternative framework that employs innovation systems perspectives on the problems of development and underdevelopment, there is a need to formulate a new conceptual framework of Innovation and Sustainable Development Systems (ISDS). This can be done by doing two main reviews: The first is to understand and develop how linkages and combinations between economic and noneconomic structure that include explicityly social and enviromental features take place. Further explorations will be needed on the following: interaction between different actors and stakeholders, how innovation and learning is embedded by economic activity generators such as firms, the link between, formal theories and appreciative theories, awareness and learning in connection between the tools used in each type of theorizing, deepening evolutionary economic dynamics to include new thematic areas such as sustainable development and national economic integration in relation to reducing dependency on donors in different types of developing and transition economies, finding productive linking internal and external, domestic and international, political and economic, social and environmental, and empirical and policy changes and approaches in different national economic settings. Combining development economics with systems of innovation in research requires suggesting the combination of innovation with sustainable development. Instead of using systems of innovation at a meta-level and try to draw or adapt it to address the dilemmas of underdevelopment, a new conceptual framework and approach can be proposed. We suggest and propose to combine development economics and the system of innovation by formulating the development economics of innovation by formulating the National Innovation and Sustainable Development System (NISDS) or the general nonregion specific focusing device that we propose as the Innovation and Sustainable Development System (ISDS). This conceptualization can link clearly development economics and systems of innovation by advancing research that combines the challenges that each side may not fully capture. It enriches both development economics and the economics of innovation that innovation system approach has promoted by strengthening the focusing device for exploring and examining the problems of development and underdevelopment with a new synthesis of ISDS.

The concept of national innovation systems (NIS) has come under close scrutiny from researchers who have witnessed the overwhelming sweep of the globalisation and internationalisation of research and innovation. Questions are being raised on whether the NIS concept is nearing its sale-by date as international and regional networks of innovation overtake them in the production of knowledge. The 'hollowing out of NIS', though not adequately researched, is an ongoing trend that any proposed review of NIS in Africa and beyond would have to address. Research has barely begun on how to harness the positive outcomes of the internationalisation of innovation systems for national development and on the policy implications for maintaining NIS in the face of sweeping internationalisation of innovation networks. Therefore, it may be useful and interesting to pay more attention to networked innovation systems with boundaries of various nature giving way to the facilitation and incentivisation at macro, meso and micro levels that can be also self-organising and self-maintaining. However, national systems of innovation remain important as laws and policies are still framed with a national context and by national governments notwithstanding the high visibility of regional and supranational networks and from macro to micro levels of systems of innovation. The national state despite the corrosion of sovereignty by globalisation, is still important and the governance and stewardship of networks of innovation rests largely on the national state or on the supra/international organisations where the role of the national state is decisive.

There are undoubtedly many areas of inquiry missed by conventional scholars of innovation studies in Africa. In Africa, the question of innovation is usually linked to economic growth and development. This is not to affirm that this trajectory is wholly undesirable or in fact over-rehearsed. However, it is also necessary to think of new disciplinary combinations and parameters that bring other fields of expertise into the discipline.

For instance, filmmakers and producers in West Africa have been able to evolve practices that follow neither the Hollywood model of film production nor Bollywood movie values. Accordingly, we have what has been termed Nollywood which has had a tremendous impact on the economy, culture and society of many regions beyond West Africa alone. Nollywood reportedly employs almost a million persons with a wide range of secondary industries attached to it. However, the Nollywood creative industry is not studied as a model of innovation and is more often confined to the realm of culture alone even though it is having a global impact in terms of transnational economies and globalizing African representation. The point to consider is that are there other innovative cultural practices within the African continent that are changing the economies and lives of Africans? Undoubtedly there are many such developments that require serious research to unearth and excavate.

This book includes research work that can potentially contribute to the emerging innovation and development field new paradigmatic combinations by adding culture. Fortunately, the subject of culture is one that is sufficiently broad to incorporate intellectuals not only from cultural studies but also literary theory, history, critical theory, social science, ethnography and anthropology. Innovation and culture is a potential sub-field under the overall emerging Innovation and Development field that began with works by Sanjaya Lall, Christopher Freeman and Martin Bell as well as early contributions in Latin America - Jorge Katz - and Asia - Linsu Kim - got consolidated and given a new momentum by the formation of the global network on innovation studies, Globelics, initiated by Bengt-Åke Lundvall in the northern developed economies in collaboration with scholars from the south (Jose Cassiolato, Mammo Muchie and Shulin Gu) and the North (Christopher Freeman, Richard Nelson and Luc Soete). (http://www.globelics.org/about-globelics/history-and-future/) We now have also journals devoted to developing this emerging field such as African Journal of Scinece, Technology, Innovation and Development (AJSTID) (http://www.tandfonline.com/toc/rajs20/current), the Journal of Innovation and Development, and the New Journal of Creativity, Innovation and Social Entreprenurship (JCISE).

In this book, we have included chapters that focus on innovation system in different sectors across different African countries. These are as following:

- Variations of innovation system from the local to the global
- Technological capability, innovation capacity and agro-industry development in Nigeria
- Determinants of technology adoption in Ugandan agricultural manufacturing firms
- Testing technology transfer policies for the South African seed industry
- Learning, upgrading and innovation in Nigerian indigenous industries
- Natural gas in future Nigerian national industrial development - an overview in the context of current progress in technology
- Uyinmwen and the development of an indigenous genre in the Edo-Benin home video industry, Nigeria
- Barriers towards the development of emerging contractors in the Limpopo province, South Africa
- Forging partnerships for innovation through networking: case studies from the Fodder Innovation Project, Nigeria
- Assessment of technological innovations in selected indigenous oilfield servicing firms in Nigeria
- Financial lending and investment in sustainable small-scale agribusiness in Sub-Saharan Africa

Baskaran and Muchie discuss variations of innovation system from the local to the global (firm level, national level, regional/local/cluster level, and sector level).

John Olatunji Adeoti and Adetola Ibidunni Adeoti discuss the strategic role of building technological capability in the Nigerian agro-industry, and examine the prospects for expanding the technological base and the innovation capacity for agro-industrial development. Case studies of technological capability in the Nigerian agro-industry are used to illustrate the critical elements in the development of technological capability in agro-industrial processing, and the role of stakeholders in partnership arrangements especially with respect to public-private sector interaction. The analysis in the paper demonstrates that the technological trajectory in industrial innovation is still at the initial conditions of articulating the frameworks for technological learning, building infrastructure for R&D, and providing adequate and appropriate incentives for industrial innovation. Consequently, the existence of technological innovation in the Nigerian agro-industry has been limited to anecdotal incidence of industrialinnovation. The discourse is concluded with specific suggestions stressing the need to integrate industrial and technology policies in an innovation policy that details specific sectoralstrategies for stimulating a dynamic industrial innovation system with a focus on agro-industry.

Niringiye Aggrey analyzes the impact of investment climate factors on technology adoption by Ugandan agricultural manufacturing firms in Chapter 4. A probit regression model was used in the analysis. Panel data for the 2000-2002 period showed that the firms most likely to adopt new technology are large, participate in export market, face stiff foreign competition, located in the main city, fast-growing, have a large share of highly skilled workers, and owned by foreign entities. On the other hand, agricultural manufacturing firms that are old, reported inadequate infrastructure as being one of the three major constraints of doing business in Uganda hence are less likely to invest in technology. Future increases in technology adoption by firms will require complementary reforms of the investment climate.

Globally, the seed industry has become increasingly concentrated in the last years. Few companies are controlling the majority of the market as well as the majority of the resources. This trend can also be observed in South Africa where multinational companies are currently controlling around 85% of the maize seed market. The aim of Chapter 5 by Sebastian Derwisch & Birgit Kopainsky is to analyze the structures that have led to the dominance of multinational companies in the South African seed industry. For this purpose, they develop a dynamic simulation model. Model simulations show market concentration as a result of the acquisition of domestic companies by multinational companies that left the South African seed industry in a locked in state. They test the effect of different technology transfer policies on the South African seed industry to see how the development of the domestic seed industry can be stimulated and how path dependency can be broken. The results of the simulation show the importance of different policies that aim to enhance spillover on the development path of the domestic seed sector. The results deliver insights for policy development for other countries in Sub Saharan Africa that are yet to develop industrial seed sectors.

In spite of the increasing globalisation of technology, the involvement of developing countries in producing new technologies and innovations is almost negligible. The production of technological knowledge is concentrated in industrialised countries and developing nations are still lagging behind. It is therefore imperative for developing countries to look inwards to deploy science, technology and innovation (STI) in specific areas where they have comparative advantage. These comparative advantages could then be developed to achieve competitive advantage. Indigenous technology is one of the veritable areas where these nations can derive global competitiveness. In Nigeria, the indigenous industries are as diverse as the culture of the people. The indigenous technologies present significant opportunities for local economic transformation and global competitiveness. This would happen if these technologies are upgraded through the deployment of STI. In Chapter 6, Siyanbola Willie O., Olamade Olumuyiwa O., Isola Oluseyi O. & Egbetokun Abiodun A. present findings from a two-year study of some major indigenous technology clusters in Nigeria.

Aminu Bayawa Muhammad, Abdullahi Abdu Zuru, Muhammad Gidado Liman & Ahmad Ahmad Tijjani discuss the natural gas sector and its impact on the future Nigerian national development in the context of current progress in technology. Nigeria is blessed with abundant natural and human resources - the basic ingredients required by any nation to reach the pinnacle of industrial development. The nation's abundant fossil fuel resource, and in particular natural gas, is a double blessing being both energy source and raw material for chemical industry. Novel industrial technologies have been, and are still being, developed for processing natural gas into cleanest fossil-based fuel and invaluable products that could further be processed into much needed clean fuels and base and intermediate chemicals for chemical industry. Thus with abundant natural gas resource, the nation could exploit these technologies and fuel her future industrial development. However, for meaningful and sustainable development, the authors argue that the technology needs to be indigenous and thus industry-academia cooperation in research and development should be well established with full support of the government.

Uyilawa Usuanlele analyses the development of an indigenous genre in the Edo-Benin home video industry in Chapter 8. The Nigerian movie industry which has been gradually developing in big screen celluloid formats, and largely as a reaction to Western stereotyping of Africa, in the early 1970s and 1980s suddenly became a big industry of world acclaim known as Nollywood in the 1990s. This was the outcome of a change to themes dealing with indigenous African cultural values and the adoption of the video format for home use given the existence of the Video Cassette Recorder and its attendant mass marketing in the 1970s. The change was also characterized by production that appealed to an indigenous constituency and adaptation to their yearnings hence the production of home videos in various Nigerian indigenous languages. The economic undercurrent for the change to the home video format has also been noted by Haynes to be borne out of the economic crisis of the 1980s. But this argument is largely based on the experience of Lagos and the Yoruba.

Lawrence K. Tshivhase and Zeleke Worku present research findings of a study that investigated factors that adversely affect sustainable development of emerging contractors in the construction industry in the Limpopo Province. The study investigated 104 emerging contractors doing construction works. The study was based on quantitative and qualitative research dissected through stratified random sampling. Results obtained from Pearson's chi-square tests of associations showed that there were significant associations between barriers to sustainable development of emerging contractors and lack of vision for business (P=0.000), lack of transparency in tender processes (P=0.000) and payment delays (P=0.009). Further, based on the results obtained from binary logistic regression analysis, growth and development of emerging contractors was significantly influenced by lack of marketing business [OR=8.58, P=0.001, 95% CI = [2.48, 29.67]] and adverse political environment [OR=15.41, P=0.002, 95% CI =[2.72, 87.34]]. According to the findings established from the in-depth interviews, emerging contractors in Limpopo are significantly impeded by lack of subcontracting works, inadequate investment in infrastructure projects and shortage of skilled personnel. From the findings of the study, key recommendations are made to the Limpopo Provincial Government and key stakeholders in the construction industry so that principal barriers deterring the development of emerging contractors can be adequately addressed.

Agricultural systems in Africa are subject to changes that defy prediction. Such changes imply that attendant challenges and opportunities do not present a defined trajectory and do not implicate a defined architecture of actors to leverage coping strategies. The general argument by Elias Madzudzo in Chapter 10 is that such complexity requires a congruence of strategies and tacit knowledge embodied not only in science and technology bases but institutional and policy realms, a departure from earlier paradigms where the former were deemed the sole answer to challenges in agricultural systems. Specifically, the paper argues that without networking, resources and expertise held by each of the actors is put to limited use. An approach that facilitates the networking of diverse sets of actors leads to opportunities for unusual partnerships that, in turn, lead to better linkage of research and development i.e. greater innovation capacity to navigate changing environments. This view and approach, it is believed, contributes to the breakdown of the barriers between science and resource users. The policy implication of this paper is that networking is a key component for the social capital of socially and ecologically resilient communities.

Oluseye O. Jegede, Abiodun A. Egbetokun and Willie O. Siyanbola examine the types of technological innovations that had occurred in indigenous oilfield servicing sub-sector within the period 2001-2010, their intensities and evaluated the factors that influenced the capability of the firms in the sub-sector. The study used primary and secondary data sources. Sixty firms were sampled. Structured questionnaire were administered in the firms. This was supplemented with shop-floor observations and interviews. The questionnaires elicited information on the firms' innovation activities and the factors that influenced these. Sixty firms were purposefully selected. A total of 60 questionnaires were administered on heads of technical, finance and administration departments in the firms out of which 66.7% were returned and found suitable for analysis. Secondary data were sourced from the internet and other published sources. Organisational innovation was the most prevalent (65%), product and process innovations were implemented, respectively, by 15% and 28.2% of the firms. Specifically, product innovations in the sub-sector majorly involved introduction of new service; and process innovations were expressed as modification of existing processes. Innovation intensity was low. The important internal factors that accounted for the innovation performance included the skill of the workforce in the technical departments and extensive staff training. The important external factors included interaction with customers, competitors, suppliers of equipment and the industry association which had been a significant path of capability building. The study found that the overall innovation of the firms for innovations was mainly expressed in organisational changes; product and process innovations in the sub-sector were mainly incremental.

Benjamin Olatunbosun presents an analysis of financial lending and investment in the small-scale agribusiness sector in Sub-Saharan Africa. Expansion of operations by rural financial and non-financial institutions in developing countries has prompted the enactment of regulations and emphasis on risk management by national supervisors. However, lending to smallholders especially sustainable framers and financial investment in small-scale climate mitigation and adaptation agriculture has not seen substantially increased. The impact of financial institutions on anti-climate-change agricultural practices and sustainable development has been advocated by a number of researchers. This study examines ways of using the opportunities provided by climate-change to reverse existing disconnect between smallholders and small-scale agribusinesses and financial institutions. Using available literature on sub-Saharan banking, climate change investment and economic development, this paper explores support mechanism for investing in climate-friendly agriculture and the possible integration of mitigation payment into Credit risk management. The collaboration between financial institutions and smallholders is an appropriate foundation for relationship building which will ease credit constraints, improve information asymmetry and increase agricultural investment.

Finally, the contributions reflect on different sectors by exploring how the dynamics of innovation is transforming the different sectors in the different countries in Africa. We invite readers to engage deeply in these impotant works to found and apply the innovation approach at the sectoral level to construct an African integrated development path that is inclusive of social and nature wellbeing.

Chapter 2

System Divergence or Coherence:

The Variations of Innovation System from the Local to the Global

Angathevar Baskaran & Mammo Muchie

Introduction

The system of innovation originated with the work of Fredrich List in 1851. In the 1930s Joseph Schumpeter developed the concept of innovation as creative destruction. In the 1980s Chris Freeman and other scholars worked on national systems of innovation. There are now varied units selected for innovation system conceptual framing from local, city, sector, nation, and region and global levels. The question is how these varied formulations of systems of innovation at different levels diverge or converge in relation to the impact measurement and output to promoting tangible development outcome.

The other dimension is the variation in the inputs for creating innovation, and the different types of innovation that emanate from incremental, radical, modular, architectural and a variety of other characteristics that generate process and product novel outputs. The range of inputs and outputs create waves of innovation that need close scrutiny given destruction and creation are embedded in the innovation dynamics.

There is now also a new concept of transformative innovation that is conncted to addressing development to become sustainable.

There are thus varied presentations of the innovation conceptual framework that have impact on how divergent or coherent the innovation path can be to generate a developmental outcome without creating major gaps and differences. If a given sector is innovative and other sectors are not or its innovative thrust is at the expense of others, this means that the system construction is defective rather than being constructive.

In countries in Africa innovation should bring integrated, transformative and sustainable development. This demand makes innovation system building from the various units from local to global to not be divergent but to be arranged in a systematic and coherent way. Something like a unified and not dispersed wave of innovation activities is needed to address all the development challenges faced by African economies currently.

Nelson and Winter (1982, pp. 128-130) have proposed boldly an innovation framework to economic theory as an alternative to neo-classical framework building on earlier criticisms of mainstream economic thinking mainly from the writings of Veblen (1909), and Schumpeter (1911, 1942) on modern dynamic economic theory building.

Issues that seem to preoccupy much of the economists hoping to create an alternative to the main-stream neo-classical economic framework appear to be understanding economic growth; short term and long term economic firm level and/or national performance, micro and meso level competitiveness, firm and national level productivity, economic catching up, learning and knowledge creation and absorption in a given economic structure, and inter linkages between firm competitiveness and national competitiveness and productivity, symmetry and system building such as national, sectoral and other types of innovation systems. Since innovation is characterised by the process of creating and destroying, some economists including Veblen earlier on have not been open to the notion of innovation systems and symmetry. They focus more on asymmetry and system breaks than makes, associating innovation more or less with a dynamic that disrupts systems and symmetry rather than the opposite.

Since 1980s theories on innovation and their use have gradually expanded their focus and complexity. From the initial focus on the individual firm or entrepreneur they expanded to include the environment and industry in which firms operate. They started focusing on the national system of regulations, institutions, human capital and government policies and programmes (Niosi et al, 1993). Subsequently, the focus also included regional level or local level systems of innovations. In other words, initial perception that innovation is basically an individual act of learning by a firm or entrepreneur has expanded to include the larger system (consisting of various institutions, policy framework, incentives etc.) in which this act occurs. It is now widely viewed and accepted that innovation is a process, which is not linear as it involves continuous interactivity between various actors and factors.

Figure 1 illustrates how the use of the concept of systems of innovation has grown and proliferated over the years. This can be traced in four major areas: (i) spatial; (ii) industry and technology specific; (iii) in terms of innovation types; (iv) in terms of level of technology/ innovation complexity; and (v) in terms of economic and social objectives. In the area of spatial we can identify global innovation systems, national innovation systems, regional and sub-regional innovation systems, and local and city innovation systems. In the area of industry and technology specific innovation systems we can see studies focused on specific industrial sectors such as manufacturing, telecommunications, automotive, agro-food and service and specific technology focused such as biotechnology, information and communication technology (ICT), and electronics. In terms of innovations types we can see the focus of studies on product innovation, process innovation, service innovation, organisational innovation and so on. Similarly, studies focused on levels of technological or innovation complexities such as incremental, revolutionary, radical, systemic and paradigm and so on. Finally, we can broadly see studies focusing on innovations driven by social objectives and economic motivations or objectives.

Figure 1: Innovation Systems - Theories/ Concepts/ Typologies/ Taxonomies

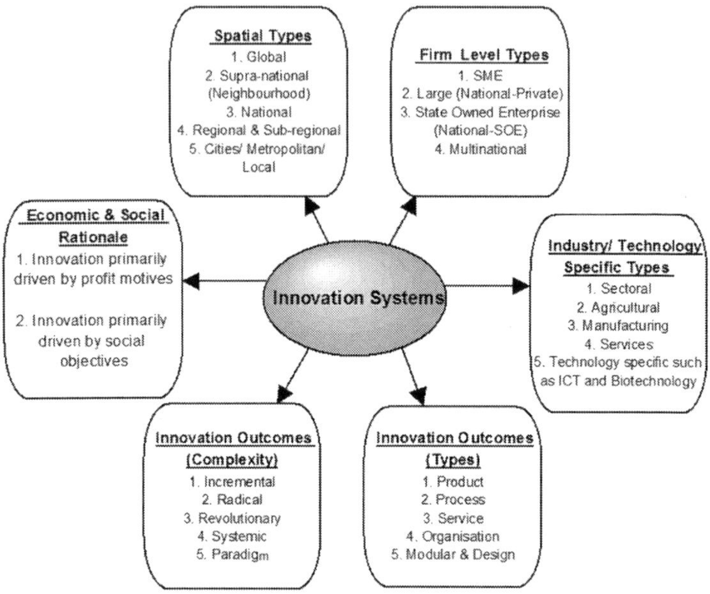

Source: Tang et al. (2015)

Theories on innovation emerged initially with the main focus on the firm and entrepreneur. Then they gradually expanded their focus to the environment and industry in which a firm operates. This led to the emergence of the national system of innovation (NIS) that includes regulations, institutions, human capital and government policy regimes. NIS framework further led to the sub-national (regional/ local/ city or metropolitan) and sectoral innovation system approaches.

Firm Level System of Innovation

Firms' internal capacity to absorb and utilize the diffusion of knowledge and generate new knowledge is important in the context of national innovation system, as firms' innovation capacity is central to a nation's industrial capacity and competitiveness (Porter, 1990). The firm level innovation may include different types such as incremental, dramatic or radical, and explicit or implicit. The innovation outcome may be due to deliberate effort by the firm or as a result of day-to-day operations.

The factors that determine the performance of innovation system at company or firm level include; (i) current technological capabilities (product, process, R&D, and skills); (ii) new product/ process development, learning; (iii) R&D investment; (iv) human resources/ capital development; (v) knowledge and information flow with customers and suppliers; and (vi) market conditions. Innovation strategies followed by firms include R&D expenditure, technology efforts concentrated on production organization and product quality and modernization of production processes, normally through equipment import, focus on learning capacity to increase organizational efficiency, the purchase of new equipment and so on. Figure 2 illustrates the firm-level innovation system/ process.

Companies/ firms play a major role in innovation process which includes internal linkages between various actors within the company or firm and external linkages, that is, with outside agents and organisations including other firms such as custom ers, suppliers, subcontractors, partner companies, financiers, research institutes, semi-public and public research and development organisations, etc. However, the degree of linkages internally and externally may be at different levels and intensity. Also the innovation could be incremental or radical or revolutionary, tactical or strategic, because of deliberate effort or just due to routine operations.

Many studies focused on the relationship between firm's innovation capability and its performance and almost all these studies found a positive and significant relationship between innovation and different measures of firm performance. For examining the relationship between innovation and firms' economic performance, several independent variables such as physical capital, human capital, R&D and other innovation-related investments as well as firm size are considered important (Janz et al., 2003).

Figure 2: Firm-level Innovation System/ Process

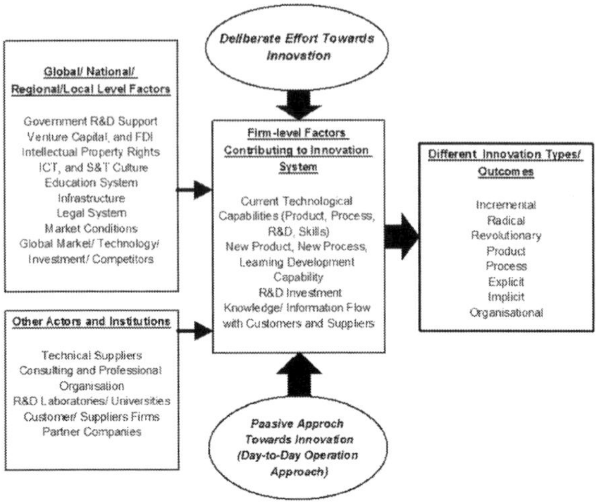

National Innovation System (NIS)

Increasingly it was recognised that innovation process is not linear, as it involved continuous interactivity between various entities such as suppliers, clients, universities, productivity centres, regulatory bodies, financial institutions and other social and economic actors (Mytelka, 2001). Also, it was recognised that innovation is not merely an individual act of learning by a firm or entrepreneur, but is situated within a larger national innovation system where a number of actors and institutions are linked and connected through flow of finance, skills, knowledge and information and is also influenced by social factors such as social rules and cultural norms.

Linear model of science and technology (Freeman, 1995), where investment in basic R&D is believed to lead to new technology applications and innovations, was increasingly found to be inadequate in explaining differential rates of technological innovation and economic development experienced by industrialised countries. That is, despite similar level of high R&D

investments across industrialized and semi-industrialized countries "evidence accumulated that the rate of technical change and of economic growth depended more on efficient diffusion ...and as much on social innovations as on technical innovations" (Freeman, 1995, p. 10). Furthermore, it was also felt that the "mainstream macroeconomic theory and policy have failed to deliver an understanding and control of the factors behind international competitiveness and economic development" (Lundvall, 2002, p. 214). Therefore, these factors led to the mergence of the concept of national innovation system (NIS).

Different authors have defined NIS in different ways. For example: Freeman (1987) defined it as the "network of institutions in the public- and private-sectors whose activities and interactions initiate, import, modify and diffuse new technologies". Lundvall (1992) defined it as the "elements and relationships which interact in the production, diffusion and use of new and economically useful knowledge... and are either located within or rooted inside the borders of a nation state". According to Nelson (1993) it is the set of institutions whose interactions determine the innovative performance of national firms.

Lundvall (2007, p. 102) argued that NSI concept can be employed at two levels: (i) the 'core' - *"firms in interaction with other firms and with the knowledge infrastructure"* including universities; and (ii) 'wider setting' that includes *"national education systems, labour markets, financial markets, intellectual property rights, competition in product markets and welfare regimes"*. In the 'wider setting' the government plays a major role in a number of ways. We would argue that in the narrow sense NSI involves a system of interaction of a wide variety of public and/ or private firms with other institutions such as universities, and government agencies -- all working together towards attaining the production and diffusion of knowledge and science, technology, and innovation within the boundaries of legally recognised states. The form of the interaction can take both technical and non-technical dimensions. It could be organisational, institutional, commercial, physical, human, mental, legal, social, and financial interactions. The broader goal of such interactions is the socio-economic development, regulation, and support for new science, technology, innovation within the country by dealing with and responding to both internal and external challenges. For this study, we employ the NSI concept in its wider setting.

Baskaran and Muchie (2010, 2010a) formulated a conceptual framework derived from the wealth of literature discussed above to investigate how and to what degree differing levels of NSI strength and weakness in different BICS economies and selected Asian economies coped in mitigating some of the adverse impacts of the recession. Baskaran and Muchie identified four key sets of elements of NSI. 1. The first set *Conceptual Framing* involves the ideas and policies that frame the overall scope or possible set of interactions of politics, economics and knowledge. The behaviour and interactions are often shaped by sets of common habits, norms, routines, established practices, rules, or laws. 2. The second set involves *Institutions, Technologies, and Knowledge and their co-evolution which enable* implementation of the conceptual framing and policies selected above (the first set) and to build an efficient innovation system. 3. The third set involves the means provided to the institutions (second set) for realising the goals set (first set), that is, various incentives such as financial and social rewards. It is vital to foster appropriate incentive system. If the incentive system is inappropriate or fails to command wider acceptance, the opportunity to organise robust NSI and achieve measurable results will be put in jeopardy. 4. The fourth set highlights the overall efficiency of the environment for learning in terms of implementation, monitoring, review, and feedback involving the above three sets. The learning outcomes can be different such as transformative, adaptive, corrective, modifying, evolutionary, and so on. This can also be negative.

Figure 3 illustrates the four sets of elements in NSI (wider setting). The strong presence and interaction and linkages between various institutions, industrial sectors, technologies, knowledge, incentives, investment, and infrastructure determine the higher or relatively stronger or weaker level of functioning of a particular NSI. We would argue that the relative strength of an NSI can have an impact on a national competitiveness and economic performance.

Figure 3: National Innovation System (Wider Setting): A Conceptual Framework

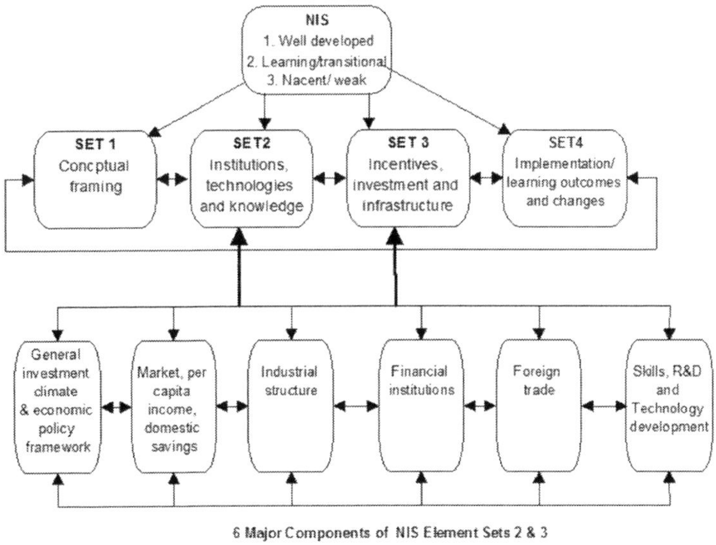

Source: Liu et al. (2011)

To recapitulate, the NIS concept recognizes that a complex relation, linkages, and co-evolution between a number of institutions also plays a major role in the innovation process apart from the firms that are directly involved in bringing new products and services to market. The actors, activities, linkages, strengths and weaknesses of national innovation system (NIS) concept are summarized in Table 1.

Table 1: National Innovation System (NIS)		
Actors/ Institutions/ Activities/ Linkages	**Strengths/ Advantages**	**Weaknesses/ Disadvantages**
1. OECD identified five institutions in NIS: (i) Governments (local, regional, national and international, with different weights by country) (ii) Bridging institutions, such as research councils and research associations; iii) Private enterprises and the research institutes they finance; (iv) Universities and related institutions that provide key knowledge and skills; (v) Other public and private organizations that play a role in the national innovation system (public laboratories, technology transfer organizations, joint research institutes, patent offices, training organizations and so on). 2. Network of these institutions in the public and private sectors interact and initiate development and production of science and technology within national borders.	1. NIS helps to study not only developed economies, but also developing economies, although there are significant socio-economic and political differences and variations. 2. NIS approach provides an alternative to neo-classical economic theories of growth. Innovation is rooted in processes of interactive learning that is problematic in pure markets. 3. NIS provides a flexible conceptual framework to study the problem of "technological gap" between the developed and developing nations (particularly the advanced and emerging developing economies). 4. NIS provides different approaches to study innovation process in developing countries such as Charles Edquist's (2001) Systems of Innovation for Development (SID) concept. Others emphasis more on learning than innovation - passive learning to absorb technological capabilities for production and active learning where deliberate effort is made to master technology.	1. In the era of globalisation the question is: how relevant is the NIS which emphasises on the national level? One can argue that there are factors that are beyond the control of national governments which can influence the innovation system. 2. International knowledge and technology flows, information and capital flows and international collaborations are increasing in volume. There are constraints imposed on nation states and NISs by: intellectual property regimes, international al trade regimes and labour systems, regional economic alliances, influence of transnational corporations on the structure of markets, and investment in R&D. 3. There is still knowledge gap in understanding fully the impact of TNCs on local learning and innovation and also the role of international institutions that shape both the strategies of firms and the policies of national governments.

3. Continuous interactivity between suppliers, clients, universities, R&D organisations, standard setting bodies, financial institutions and other critical social and economic actors. 4. Interaction among these units may be technical, commercial, legal, social, and financial with the main goal of the development, protection, financing or regulation of new science and technology.	5. Despite the inevitable question as to the appropriateness of the concept of NIS due to the emergence of globalized economy, it is widely accepted that domestic policies, actors and institutions still play an important role. 6. NIS also helps to capture the uneven economic development dynamics in developing economies. Also, it helps to identify the linkages between innovation systems and industrial economic narrowly, and more broadly structural social and economic development/transformation. NIS model also enables elaboration and variations to analyse and capture particular aspects of innovation process (e.g. NIS's impact on FDI, and FDI in R & D).	4. There are still problems in applying the NIS concept to large number of small and least developed economies in Asia, Africa, and Latin America, as the institutions and actors that are central to NIS are hardly present in these economies. 5. Another fundamental problem with NIS is determining its scope within the national boundary. That is, an analytical distinction between a "narrow" NIS concept, and a "broad" NIS perspective. 6. Although there are strong measures such as R&D expenditure and patents, there are still problems in measuring the linkages, flows and outcomes in the NIS.

Source: Drawn from literature by the authors

Cluster/ Regional/ Local/ City (Metropolitan) innovation system

Increasingly, researchers are analysing innovation systems by taking sub-national level entities, as it is felt that NIS is inappropriate or inadequate for such studies. These include concepts such as 'cluster', 'regional' 'city' 'metropolitan' and 'local' innovation systems. Often these are not distinct from each other. For example, review of writings focusing on sub-national level approach reveals that 'cluster' and 'regional' are used interchangeably to mean the same thing and 'regional', 'local', 'metropolitan' and 'city' innovation systems are used interchangeably to mean the same thing.

The 'cluster' approach to innovation (as illustrated by Figure 4) was introduced by Porter by emphasising the importance of firm interactions with supply chains and with public research organisations (Porter, 1990; 1998). This gained wider acceptance among policy makers around the world because of the emergence of successful industrial clusters or regions such as the Silicon Valley in the US and similar regions in Italy and Germany. Consequently, the 'region' became an alternative level of analysis (Acs and Varga, 2002; Cooke, 1992). Regional innovation system can be defined as a system of innovation that strengthens the competitiveness of a region through strengthening the capabilities of firms which in turn can lead to the growth of the region (Cooke and Morgan, 1998).

Figure 4: Innovation System - The Concept of Cluster

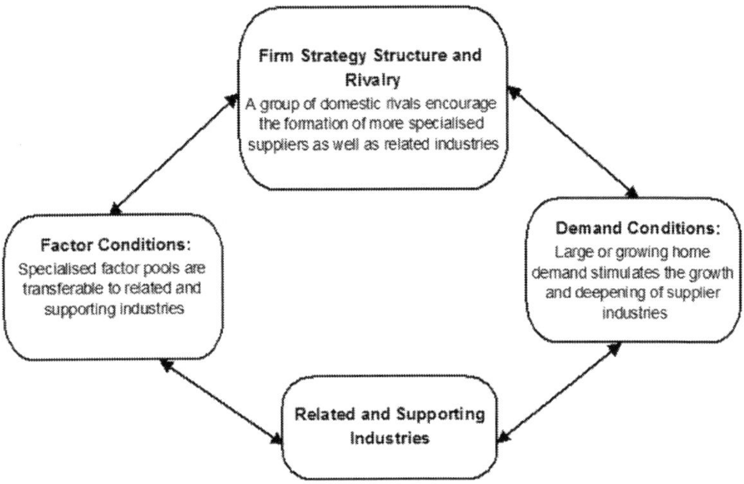

Source: OECD (1997), *National Innovation System*, p. 26 (Original source: Porter, 1992)

Figure 5: Regional/ Metropolitan Innovation System - A Conceptual Framework

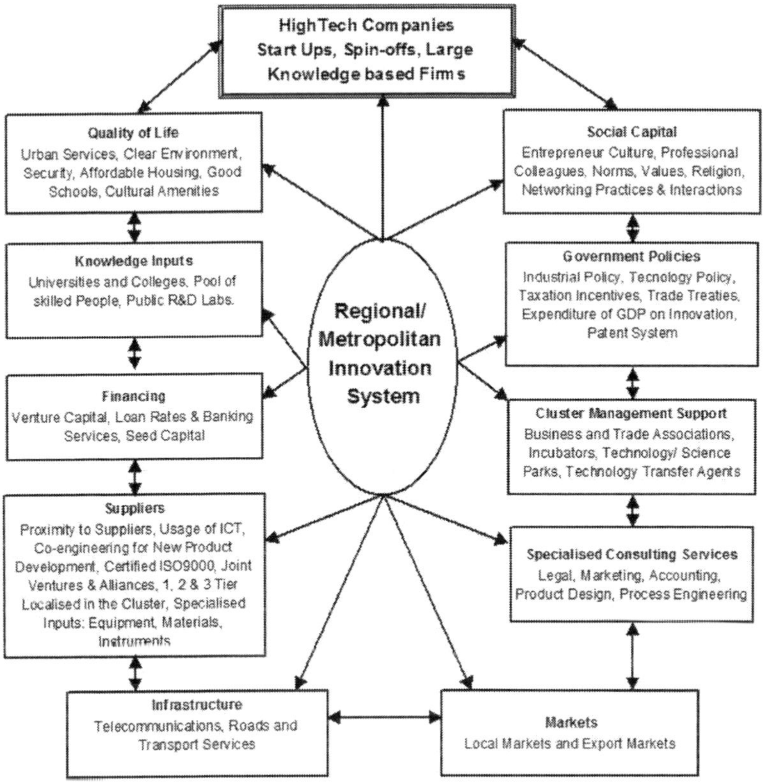

Source: Flores, M. (2004)

Increasingly local innovation networks are considered to play an important role in the innovation process and economic growth of regions and cities. Local/ regional innovation process results due to interactions at economic and social levels between different institutions located in a particular region. Therefore, to understand the efficiency or performance of a regional/ local innovation system, it is important to examine not only the horizontal and vertical relations among firms but also the linkages between firms and other institutions such as universities, research institutions, supporting industry, provincial/ local government policies, and financial institutions. In other words, it is considered that the synergy among various institutions and actors in a region plays an important role in creating an efficient regional innovation system by increasing its innovation capability.

However, increasingly it is perceived that the conditions of regions have changed significantly due to globalisation -- emerging global market and value chain, global competitors, global finance and more networking structure (Gereffi 1999; Schmitz 2004; Flores, 2004).

Figure 5 illustrates different actors, activities of regional innovation system and linkages between them. Table 2 presents actors, activities, linkages between them, strengths and weaknesses of the Regional/ Local innovation system.

Table 2: Regional/ Local/ City/ Metropolitan Innovation System

Actors/ Institutions/ Activities/ Linkages	Strengths/ Advantages	Weaknesses/Disadvantages
1. Local innovation networks are considered to play an important role in the innovation process and economic growth of regions and cities. 2. Local/ regional innovation process results due to interactions at economic and social levels between different institutions located in a particular region.	1. Nation innovation system does not fully explain why there is an uneven development across different regions in a country. Regional innovation system concept helps to address this problem. It helps to understand the gap between advanced regions and less advanced regions. 2. It provided an alternative to growth without employment at national level. 3. It emphasises more on networking and linkages among regional actors and institutions which helps to understand better problems associated with regional/ local development.	1. Due to differences across regions within national boundaries and across different countries it is difficult to device 'common solutions' or general 'best practices' for regions. This creates problems towards policy formulations and learning from other regions. 2. Due to differences across regions, one of the major problems with regional innovation systems is the lack of comparability across regions. It seems every region aims to develop an innovation system based on some successful models which may not be appropriate for them.

3. To understand the efficiency or performance of a regional/ local innovation system, it is important to examine not only the horizontal and vertical relations among firms but also the linkages between firms and other institutions such as universities, research institutions, supporting industry, provincial/ local government policies, and financial institutions.	4. Regional innovation system approach helps to understand to some extent why many of the leading firms in 'new economy' industries which need to innovate at a rapid pace have emerged in the same few locations across the world. 5. Regional innovation system helps to understand not only the region-level interaction to promote innovation in traditional industrial sectors such as manufacturing, but also in service-related sectors. 6. Analysis at sub-national innovation system level helps to understand the factors that help sustainable competitiveness of SMEs and their technological capacity-building.	3. The concept of regional innovation system can lead to confusion, as 'region' is interpreted in different ways (global region, or supranational region, metropolitan or city region, sub-nation region or local). 4. Although it can be argued that innovation is strongly influenced by region-specific factors, the ability of and incentives for firms to innovate are mainly linked to national level factors such as intellectual property right laws, taxation, corporate governance, tariffs and so on. Also by global factors such as increasing competition in the global market.

Source. Drawn from literature by the authors

Sectoral Innovation System (SIS)

The concept of sectoral innovation system (SIS) attempts to provide a multidimensional, integrated and dynamic view of sectors and helps to analyse sectors which allow for comparability (Malerba, 2002). The sectoral system of innovation approach encompasses and includes the technological system approach, by placing it within the sectoral context (e.g. biotechnology, ICT, aerospace, nanotechnologies, electronics, telecommunications, pharmaceuticals, automotive, and energy).

Figure 6: Sectoral Innovation System (SIS) - A Conceptual Framework

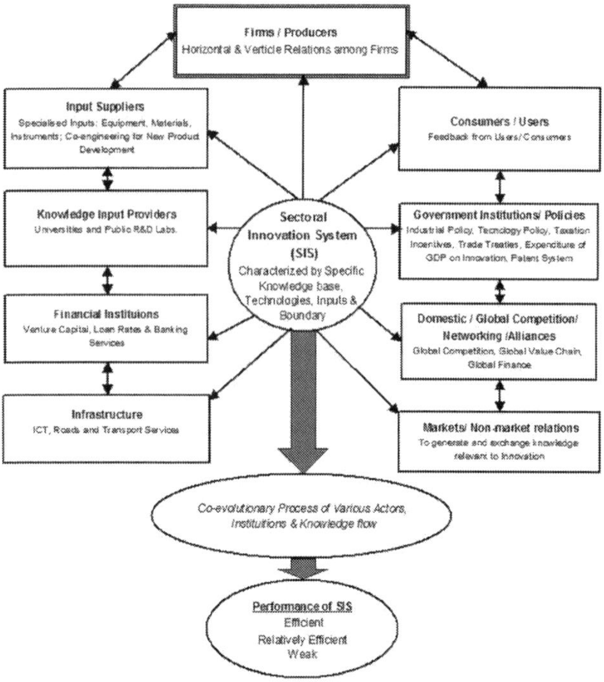

The sectoral or technology specific approach focuses on three major areas that influence the technological capabilities at the sectoral level or in specific technology area: (i) knowledge and information flow; (ii) actors and networks, institutions; (iii) linkages between these entities. This is illustrated by Figure 6.

A sector is characterised by a specific knowledge base, technologies, inputs, and a boundary which could change over time. Also linkages between various entities and activities play a major role in defining the real boundaries of a sectoral innovation system. These linkages can be static as input-output links and dynamic which take into account complex interdependent relations. Particularly, dynamic linkages among various actors and activities play a major role in bringing about technological change and growth in a particular sector. A sector is composed by various agents and organisations such as consumers, entrepreneurs, users, producers and input suppliers, firms, universities, financial institutions, and government institutions. These actors and organisations interact through both market and non-market relationships to generate and exchange knowledge relevant to innovation and its commercialisation. However, the nature of relationships and networks differ across different sectoral systems. Furthermore, it is likely that during the evolution of sectoral systems the technological and learning regimes will experience changes. Such change is also likely to result in a co-evolutionary process of various actors, institutions and knowledge flow (Malerba, 2002).

Advantages/ Strengths of Sectoral Innovation System

(i) The concept of sectoral innovation system provides an alternative analytical framework to the traditional concept of sector used in industrial economics, as it helps analyse other agents in addition to firms and brings in to focus the importance of knowledge flows, boundaries, market and non market factors and their interactions, and different institutions.

(ii) Also, this approach recognises that firms play active role in shaping their technological and market environments, unlike the traditional industrial economic view that they are passive, that is, they transform inputs into outputs in response to market price signals (Malerba, 2002).

(iii) The concept of sectoral innovation system is useful to: (a) analyse the differences and similarities in the structure, organisation and boundaries of sectors; (b) understand the differences and similarities in the working, dynamics and transformation of sectors; (c) identify the factors affecting innovation, commercial performance and international competitiveness of firms and countries across different sectors; and (iv) for the development of public policy initiatives (Malerba, 2002).

(iv) The sectoral system of innovation concept complements other concepts within the innovation system literature such as national systems of innovation delimited by national boundaries, and regional/local innovation systems in which the boundary is the region. National/ regional boundaries are not always the most appropriate ones for an examination of the structure, agents and dynamics of linkages in a sector.

(vi) Sectoral system of innovation approach helps to understand why often the characteristics of national institutions favour specific sectors that largely reflect the characteristics of these institutions. That is, some sectoral systems become far more important in a national economy compared to others.

(vii) Relationship between national institutions and sectoral systems can be a two-way street. That is, although often national institutions impact on sectoral systems, sometimes it occurs in the other direction as well. For example, the institutions of a sector may become national when their contribution (employment, competitiveness, and so on) becomes very important at national level (Malerba, 2002).

(viii) Emphasis on the diversity of sectoral systems helps to formulate different policy measures for different sectors.

Weaknesses/ Disadvantages of Sectoral Innovation System

(i) Interactions between various agents in the sectoral system of innovation are shaped by institutions at both sectoral and national levels. Many institutions such as patent system are national. This shows that it is not easy to distinguish the boundary between national and sectoral. Furthermore, the characteristics of these institutions (norms, routines, common habits, established practices, rules, laws, standards) at both levels are nearly indistinguishable.

(ii) Similarly, sectoral innovation systems are also shaped by institutions at global level. In some cases, the relevant geographical boundaries are global as well as sectoral. In such cases it is not easy to distinguish the boundary between global and sectoral.

(iii) The relationship between national institutions and sectoral systems could be different in different countries. That is, the same institution may take different features in different countries, and thus may affect the same sectoral system differently in different countries.

(iv) The nature of relationships and networks differ across sectoral systems and therefore it can be difficult and complex to compare them to each other.

Conclusion

Regardless of whether system building or not occurs in real economic systems, the national system of innovation perspective has been popularised. Its main inquiry is to understand the variations or differences in the innovation performance of nations that enters into explaining the long-term economic performance, national productivity measured in such macro-economic variables as GDP and national competitiveness. The degree to which micro-level firm innovative capability, performance and competiveness can be aggregated to contribute to national innovative productivity, performance and competition has been analytically contentious.

However, the discussion of distinct and different systems of innovation – national, regional/ local/city, sectoral/technology specific clearly illustrates a number of common characteristics: (i) they consist of a network of actors and agents together with the institutions and policies that influence their innovative behavior and performance; (ii) the presence of an interactive process in which firms/enterprises interact with each other to innovate, develop and commercialise new products and new processes and bring abvout new forms of organization; (iv) Role played by the institutions and organizations such as universities, public R&D organisations, regulatory and standard setting bodies, specialist service provders, banking and other financing institutions in facilitating and supporting this process; (iv) some major aspects of system of innovation at any boundary level: knowledge and information flow; actors and networks, institutions; linkages between these entities; investment and learning; (vii) The analytical framework (based on evolutionary theory) for systems of innovations places main emphasis on dynamics, process and transformation; (viii) In different innovation systems, the learning, behaviour and capabilities of agents are determined by the technology, knowledge base and institutional context in which firms act; (ix) Whether it is at the firm-level, or regional-level, or national level, the common view is that innovation is the engine of growth.

When we compare the actors, activities and linkages between different types of innovation systems and also their strengths and weaknesses, it is clear that although there are some clear differences in the characteristics and emphasises among different types of innovation systems, there are also a number of common characteristics among them. There is still the issue of a system coherence within a given sector and nits relationship with the national system of innovation. Serious research is needed to link the divergence and coherence relationships between the system of innovation at the local, sectoral, regional and national levels. A unifed system of innovation is waiting to be made addressing the problems of divergence or coherence and the impactin bringing tangible and measurable outcome.

References

Acs Z. J. and Varga A. (2002), "Introduction to the special issue on regional innovation systems," *International Regional Science Review*, vol. 25, no. 1, pp. 3-7.

Baskaran, A. and Muchie, M. (2010), 'Can the Relative Strength of the National Systems of Innovation Mitigate the Severity of the Global Recession on the BRICS?' *IERI Working Paper 2010-001*, Tshwane University of Technology, Pretoria, South Africa.

Baskaran, A. and Muchie, M. (2010a), 'Mitigating Impact of National Systems of Innovation on the Severity of the Global Recession: Comparison of Selected Asian Economies', *DIIPER Research Series*, Aalborg University, Denmark.

Cooke P. (1992), "Regional Innovation Systems - Competitive Regulation in the New Europe," *Geoforum*, vol. 23, no. 3, pp. 365-382.

Cooke P. and Morgan K. (1998), *The Associational Economy: Firms, Regions, and Innovation* (Oxford.: Oxford University Press).

Flores, M (2004), 'Proximity and Learning in Metropolitan Innovation Systems, Towards the Formation of High TechClusters in Monterrey and Milan,' (Research Proposal), Globelics Academy, Lisbon.

Freeman C. (1987), *Technology Policy and Economic Performance: Lessons from Japan* (London: Pinter).

Freeman C. (1995) "The 'National System of Innovation' in Historical Perspective," *Cambridge Journal of Economics*, vol. 19, no.1.

Gereffi G (1999), "International trade and industrial upgrading in the apparel commodity chain," *Journal of International Economics*, vol. 48, pp. 37-70

Janz N., Loof H and Peters B. (2003), "Firm Level Innovation and Productivity–Is there a Common Story Across Countries?," Discussion Paper No. 03-26. See *<ftp://ftp.zew.de/pub/zew-docs/dp/dp0326.pdf>*

Liu, J., Baskaran, A. and Muchie, M. (2011), Global Recession and the National System of Innovation in China: 'A Blessing in Disguise'? *African Journal of Science, Technology, Innovation & Development*, Vol. 3, No. 4. pp. 107-140.

Lundvall B. A. (ed.) (1992), *National Innovation Systems: Towards a theory of innovation and interactive learning* (London: Pinter).

Lundvall B. A. (2002), "Towards a learning society," in Conceicao P., Heitor M., and Lundvall BA., (eds.), *Innovation, Competence Building and Social Cohesion in Europe: Towards a Learning Society*, (Cheltenham, UK: Edward Elgar).

Lundvall, B.-Å. (2007), 'National Innovation Systems-Analytical Concept and Development Tool', in Industry *and Innovation*, 14 (1): 95-119.

Malerba, F. (2002), "New challenges for sectoral systems of innovation in Europe," DRUID *Summer Conference 2002: Industrial Dynamics of the New and Old Economy -who is embracing whom?*, Copenhagen, Denmark, June 6-8.

Mytelka L. (2001), "Promoting scientific and technological knowledge for sustainable development," *Paper for the Third UN Conference on Least Developed Countries,* Round Table: "Education for All and Sustainable Development in LDCs," May.

Nelson R. R, (1993) National Innovation Systems: a comparative study (Oxford: Oxford University Press)

Nelson R. Winter S. (1982), *An evolutionary theory of economic change* (Cambridge, MA: The Belknapp Press of Harvard University Press).

Niosi J., Saviotti P., Bellon B., and Crow M. (1993), "National systems of innovation: In search of a workable concept," *Technology in Society,* vol. 15.

OECD (1997), *National Innovation* Systems (Paris: OECD). See < http://www.oecd.org/dataoecd/35/56/2101733.pdf>

Porter M.A. (1990), *Competitiveness of Nations* (Cambridge: Cambridge University Press).

Porter M. (1998), "Clusters and the new economics of competition," *Harvard Business Review,* vol. 76, no. 6, pp.77-90

Schmitz H. (ed) (2004), *Local Enterprises in the Global Economy* (Cheltenham, UK: Edward Elgar).

Schumpeter, J.A. (1934), The theory of Economic Development: An Inquiry into Profits, Capital, Credit, Interest and the Business Cycle, London: Oxford University Press.

Schumpeter, J.A (1942) Capitalism, Socialism and Democracy, London, Allen and Unwin.

Tang, M., Baskaran, A., Yan, H. & Muchie, M. (2015), 'Strengthening regional integration/cooperation with the Neighbourhood System of Innovation conceptual framework: the case of China and ASEAN', *Asian Journal of Technology Innovation,* 23:2, 205-229, DOI: 10.1080/19761597.2015.1074511

Veblen, T, (1919) The Place of Science in Modern Civilisation and Other Essays, New York, Augustus M. Kelley

Chapter 3

Technological Capability, Innovation Capacity and Agro-industry Development in Nigeria

John Olatunji Adeoti & Adetola Ibidunni Adeoti

1. Introduction

Strengthening technological capability and innovation capacity are pursuits of developing countries that have achieved significant reduction in poverty and are on the path to technological and economic catch-up. Technological capability is a major determinant of innovation capacity. Though efforts and resources have been committed to building local technological capability in Nigeria, significant impact on industrialization and other economic activities is yet to be apparent. In other words, the innovation capacity is weak, and consequently, opportunities for growth and development are limited in critical sectors of the economy. Boosting indigenous technological capability thus remains a major development challenge in Nigeria. The challenge is more pronounced in the industrial sector because industry forms the centre of technological activities in any economy. In a national system of innovation (NSI), firms, which constitute the industry, are the centres of technological innovation that sustain economic growth (Freeman, 1995; Kim, 1997; Muchie and Baskaran, 2009). It has been severally demonstrated from economic theory and empirical studies that technological innovation is the engine of economic growth (Freeman, 1995; Freeman and Soete, 1997; Goel et al., 2004). Technological innovation is however not limited to the secondary sector of the economy. Technological innovation as engine of economic growth cuts across every sector. Science, technology and innovation in the primary and tertiary sectors affect the economy as much as innovation in the secondary sector. However, for a late industrialising country such as Nigeria, the transition from dependence on the primary sector to the secondary and tertiary sectors has been delayed because of the relatively poor local technological capability. Nigeria's indigenous technological capability is not only limited in its functions but also lack significant capacity to assimilate, absorb or adapt foreign technology (Okejiri, 2000; Adeoti, 2002a). The agricultural economy in Nigeria is largely subsistence and efforts at building capacity for innovation are yet to achieve significant results. A recent study by Adeoti et al. (2010) disclosed that tackling innovation deficit in the agro-food processing in Nigeria remains difficult in spite of improved contribution of agriculture to GDP in recent years. The empirical evidence from the analysis by Adeoti et al. demonstrated that building innovation capacity in the Nigeria agro-industry is constrained by poor infrastructure, inadequate policy, and inappropriate attitude of key innovation agents. Consequently, there has been limited firm-level R&D capability, and knowledge institutions such as the universities have been unable to achieve commercialization of inventions that could engender innovation in agro-industry.

In this chapter we will discuss the strategic role of building technological capability in the Nigerian agro-industry, and examine the prospects for expanding the technological base, the innovation capacity and human capacities for agro-industrial development. We will also discuss the scope for adapting and upgrading technology used for processing traditional/indigenous products, and consider the specific roles of public-private sector interaction.

The next section briefly discuss the concepts of technology and technological capability; section three presents an overview of the state of industrial development in Nigeria; section four examines the role of indigenous technology in agro-industry; section five presents case studies of technological capability in the Nigerian agro-industry to illustrate the critical elements in the development of technological capability in agro-industrial processing, and the role of stakeholders in partnership arrangements especially with respect to public-private sector interaction; and the final section concludes the paper with discussion of the prospects for improving the technological base and innovation capacity in agro-industrial development in Nigeria.

2. Concepts of Technology and Technological Capability

Having substantially solved the riddle of the so-called 'black box' of technology, neo-Schumpeterian economists generally acknowledge technology as the application of scientific knowledge to the setting up, operating, improving and expanding of productive facilities (Lall, 1987). This notwithstanding, based on the exposition by Clark et al. (1988), there are three acknowledged concepts of technology that are of interest to the discourse In this chapter. First, technology may be appreciated as physical artefacts that could be employed for the achievement of specific objectives. In this respect, technology has been considered as 'equipment', 'tools' or 'hardware'. Secondly, technology could be regarded as a flow process involving an array of these artefacts linked together according to a predetermined set of norms or standards for achieving productive objectives. From this perspective, technology has been defined as 'process lay-out', 'workflow process' or 'physical workflow'. Thirdly, technology has also been defined as 'pattern of operations' or 'body of ideas, which express the goals of the work'. Furthermore, Clark et al. pointed out that existing studies on implication of new technologies for work organisations show that technology could also in an organisational sense refer to social arrangements such as factories or bureaucracies which has been created to achieve specific planned objectives. Our view of technology In this chapter would exclude technology as an organisation apparently because we see organisation as distinct from technology, and regard organisation as the unit where the technical change process or innovation mostly occurs.

Lall (1987) pointed out that technological capability is the mastery of a broad range of technological activities. Such activities include pre-investment choice, project execution, plant operation, technological improvement, and technological transfer. This appears to be an all embracing definition of technological capability. However, the issues involved are not as simple. Technological capability is dynamic by nature and there are peculiarities to building technological capability in different sectors of the economy. Stewart (1983) showed that technology, and by implication, technological capability consists of techniques. These techniques are however not always codified or learnable through some 'blue print'. They may be 'intangible' or idiosyncratic by nature. This is largely the reason why Nelson and Winter (1982) and Freeman and Soete (1997) affirmed that technological capability is acquired from experience, conscious effort in monitoring and keeping track of developments throughout the world, accumulation of added skills, and by responding to new pressures and opportunities.

It is pertinent to point out that technological transfer or acquisition is often discussed without much consciousness for what the two terms connote. Technological acquisition is the converse of technological transfer on the giver-receiver basis. However, from economic history there has been no case of a nation transferring her technology to another, except in cases when such technology has become somewhat obsolete or common knowledge with no significant intellectual property barrier. The experience has always been a nation acquiring the technology of another nation that is at a higher frontier of technological development. Moreover, the new regime of intellectual property protection beginning from the decade of the 1980s make it imperative for developing countries to think less of technological transfer, but rather technological acquisition.

3. Agro-industry and the State of Industrial Development in Nigeria

Industrial development is arguably the engine of sustained long-term economic development. Industrial development represents a deliberate and sustained application and combination of suitable technology, human resources, management techniques and other resources to improve the production system. The economic argument for embarking on industrial development is that rising productivity in the economy depends largely on industrial performance often obtained through technological innovation, which result from improved factor input combinations. The higher wage rate associated with the modern industrial sector attracts the surplus labour from the subsistence traditional sectors. In countries where technological innovations in both the secondary and primary sectors have advanced, industrial development has produced synergy between the two sectors, and net economic welfare has improved considerably for every strata of the population.

Generally speaking, the main features of industrial development in Nigeria may be enunciated as follows:

- Low capacity utilisation by firms: Capacity utilisation which was as high as 70% in the 1970s declined to 32% in the 1990s and is presently reported to be 44.3% in 2002 and 46.2% in 2003.
- Low contribution to Gross Domestic Product (GDP). The contribution of industrial sector has been only about 4% in recent years. Low industry value addition and overtly import-dependent production system.
- Low contribution of small and medium-sized enterprises (SMEs) to total industrial production, and weak linkage between SMEs and large-sized firms.
- Absence on local industrial R&D: industry is largely user of results of foreign R&D. Multinational affiliates are notably dependent on their parent companies for R&D while local firms have little interest in investing in R&D. Consequently, there is lack of locally based technology-intensive firms.

- Low employment in the industrial sector arising from industrial closures due to relatively high cost manufacturing environment. For example, the textile sector, which used to be the highest single employer of labour in the private sector in the 1970s, has virtually wound up. Efforts to revive the sector remained unsuccessful.
- Limited exports of industrial goods (less than 1% of total exports).

Based on the above characteristics, it is pertinent to state that industrial development in Nigeria shows an imbalance in the national capacity to produce and consume technology. On the production side, the economy is predominantly agricultural or mining-based; has a relatively small manufacturing sector; a trade structure dominated by exports of primary products and the importation of manufactured goods; a low level of literacy, and a deficiency in skilled personnel. On the consumption side, the domestic market though large, has relatively weak purchasing power; low per capita income; majority of the population relied on subsistence production; and consumption patterns is skewed towards the urban consumer whose tastes are fashioned after imported goods. In these circumstances, agro-industry remains highly constrained, and policy mechanisms that promote an effective national system of innovation would be required to build innovation capacity in agro-industry (Adeoti and Olubamiwa, 2009; Adeoti et al., 2010).

As reported by Ekundare (1973), Nigeria's formal attempt at industrialisation began in pre-independence times with the Aid to Pioneer Industries Ordinance of 1952. The starting point was largely promotion of cottage industries. Later, on a more elaborate basis, import substitution industrialisation (ISI) strategy became dominant in the 1960s to 1970s. This gave way to the export promotion initiatives and deregulation of the 1980s and 1990s because of the apparent inability of the ISI to stimulate structural transformation (Adeoti, 2002b). The decades of the 1960s and 1970s experienced a relative boom in industrial production heavily subsidised by the oil economy which experienced a boom during the period. The decade of the 1980s and 1990s were largely period of de-industrialisation in Nigeria and many other countries in sub-Saharan Africa (Jalilian et al., 2000). The World Bank/IMF economic structural adjustment programme (SAP) arising from the so-called Washington consensus was the dominant reform instrument in these countries during the period.[1] One of the main objectives of SAP was to stimulate the productive base which was mainly agricultural production and the related agro-industry. For Nigeria, like in many other countries, the impact of SAP was controversial. While there was some improvement in production of export commodities, particularly cocoa beans, there was no evidence of improved agro-industrial processing. Rather, even the cocoa export later declined due to world commodity price crises, and the peculiar situation in Nigeria whereby cocoa export served as an avenue for capital flight by businessmen and even industrialists that were scared of the protracted economic crises.

The decade of the 2000s has however brought about economic reform which is able to make significant impact on agro-industrial processing. Adeoti and Olubamiwa (2009) provided the illustration of the impact of presidential cocoa re-birth initiative on innovation capacity building in cocoa production and processing. While farmers have benefited from government distribution of improved cocoa seedlings, the few existing cocoa processing firms are actively involved in exploiting the opportunities created by the initiative to improve their innovative performance. The Presidential Initiative on Cassava Processing and Export (PICPE) has also been shown to have significant impact in encouraging technological innovation and the mediation of interactions between public and private sector agents. Furthermore, Adeoti et al. (2010) provides detailed analysis of how PICPE enabled the local design and fabrication of the first medium-sized cassava flash dryer in a collaboration involving Nigerian scientists/engineers from public universities and research institutes, a private firm, Raw Material Research and Development Council and the International Institute of Tropical Agriculture (IITA). Adeoti et al. (2010) argued that though the cassava flash dryer success story may be regarded as an anecdotal evidence, it suffices that such innovative outcomes portends glimpses of hope for agro-industrial innovation in Nigeria's innovation-deficit economic landscape. Moreover, previous policy impact of the ban on imported barley malt in the late 1980s provided a series of agro-industrial innovations in the Nigerian brewing industry, where local sorghum was effectively substituted for barley malt in beer production (Ogun, 1995). It thus appears that there is a critical role for government as catalyst for building innovation capacity in the Nigerian agro-industry.

4. The Role of Indigenous Technology in Agro-Industry

At a meeting[2] of science and technology experts on the promotion of indigenous technology for development in Africa, a working definition for indigenous technology emerged as follows:

Indigenous technologies consist of all the skills, techniques or know-how that are traditional/homegrown or generated in any given location, community, country or region. It may be regarded as a component of the indigenous culture or tradition, and belief in a particular locality. Most indigenous or traditional technologies were generated, invented, or evolved by trial and error or intuition over long periods of time.

Several authors (e.g. Okigbo, 1996; Maduemezia, 1996) have advocated that indigenous technologies and modern technologies should complement one another, and this is often achieved through a process of technological upgrading in industry. In many countries that have tackled underdevelopment and achieved significant industrialisation, indigenous technologies developed alongside modern technologies. Many technologies were developed before the advent of modern science and technology, but modern technologies supported and upgraded them. For example, the fountain pen was invented before rubber and plastics were invented, yet the latter invention revolutionised the former, and added a new product called biro pen.

Specific examples of indigenous technologies on which modern industries could be based in Nigeria exist in agriculture, medicine and capital goods sectors of the economy. It is particularly noteworthy that agro-industry could be an important feature of virtually every sector of the economy through value addition and inter-linkages of critical value chains. In this respect agro-industry has specific relevance in sectors such as chemicals and pharmaceuticals, consumer food products, industrial intermediate goods, textiles, post-harvest technology hardware, etc.

It is also important to note that indigenous technologies complemented by modern technologies are critical for the development of small and medium-sized enterprises (SMEs). In developing countries agro-industry thrives mainly as SMEs. SMEs, particularly those in the manufacturing sector, play significant roles in the utilisation of indigenous technologies and technological upgrading in agro-industrial subsector. The importance of SMEs for the Nigerian agro-industry is premised on the following factors:

- labour intensive content;
- local capital (finance) involvement instead of foreign exchange expenditure;
- local physical capital inputs (locally fabricated machines and equipment) instead of imported machines and equipment;
- relatively high level of use of locally sourced raw materials;
- relatively cheap and standard quality products that is within the purchasing power of most consumers;
- relative ease of entrance into production by newcomers; and
- simpler infrastructure requirements.

Promoting the development of indigenous technologies in the Nigerian agro-industry is therefore an important instrument of SME development, which may consequently lead to the generation of employment[3] and income as well as the extension of industrial production over a greatly diversified base. In this context, the development of SMEs in the Nigerian agro-industry should accord more priority to projects that can create forward and backward linkages with the rest of the economy, promotion of export-oriented industries that utilise large volume of local raw materials, technology and equipment.

5. Case Studies of Technological Capability in the Nigerian Agro- industry

The state of local technological capability in Nigeria's agro-industry can be illustrated by the case of one engineering firm and two agro-food processing firms studied by NISER.[4] The selected three firms are SMEs, and the two agro-allied firms produce both intermediate and consumer goods. This selection is premised on the notion that the engineering sector is critical to technological capability development in manufacturing technology; and the agro-industrial sector is a major subsector that has potential for taking advantage of mastery of relatively unsophisticated technology to achieve significant progress in manufacturing. The discussion of the case studies will focus on three key areas that should highlight the critical elements in the development of technological capability in agro-industrial processing, and role of stakeholders in partnership arrangements especially with respect to public-private sector partnership. The case studies, though limited in their scope, also provides insights into the interactions between indigenous and modern technologies in the Nigerian agro-industry subsector. The analysis would focus on the sourcing of input materials; manufacturing technology; and market for products.

Case Study 1: The engineering firm

Establishment and products

The firm is an engineering firm established in 1979, and is 100% Nigerian owned. Its products include vehicles (currently produce tricycles); food and agricultural products processing machines (pepper grinders, corn shellers, rice hullers, pounding machine, etc.). The firm employs about 50 people currently out of which are 2 engineers.

Sourcing of input materials

Virtually all the inputs into the firm are imported. Sheet metal which is a major raw material for the firm is currently not produced locally. The machines and equipment are all imported but the firm has gained considerable experience in maintenance over the years. While the firm was established in 1979, the machines and equipment, which were installed in the early 1980s, are still the ones being used at the time of this study. Some of the machines though obsolete by international standards, are among the best available in Nigeria. The firm considers these machines "too sophisticated" for the Nigerian operators due to the relatively low skill levels.

The firm explained that no economic policy reform in Nigeria has so far significantly affected the production of critical engineering skills. The firm reckons that there is actually no practice of engineering in Nigeria, except for civil engineering. Major skills upgrading in the firm has been due to the firm's own initiatives. For example, about 14 years ago one of the engineers was sent by the firm to the United Kingdom to learn dye casting. Most of the skilled staff being used are those trained by the firm about 20 years ago. This is an indication of the technological backwardness in the Nigerian engineering sector. The only place where there exist better machines and equipment is the Nigerian Machine Tools Limited (NMTL), Osogbo. However, the facilities at the NMTL are not accessible to local entrepreneurs. Besides, the public sector is unable to employ it for significant production activity.

An important area where economic policy reform could have affected the sourcing of input materials is the tariff regime. Apart from the fact that the engineering sector is not taken into consideration in recent policy such as the ECOWAS Common External Tariff (CET), the implementation of any policy that relate to importation in Nigeria is problematic. Goods that should serve as critical production inputs are held up at the ports sometimes for two months or more.

Manufacturing technology

Generally speaking, technology and industrial policies have not had any significant impact on the firm's manufacturing activities. Old problems of neglect by government and policies that do not promote technological upgrading and industrialisation persist.

An example was given of a pepper grinder. Scrap aluminium from old vehicles was used because import of aluminium is too expensive. Other options are:

- take the design to China or Japan, bring home the parts and assemble the grinder;
- import the fully built grinder;
- make some components locally and order some from abroad. This is constrained by lack of equipment that can make quality and quantity of local production to match quality and efficient scale of operation of the foreign produced components. For example, there is no computer-aided manufacturing (automatic machines) in the engineering sector in Nigeria. The automatic machines available at NMTL are not accessible due to policy failure. Besides, there is no culture of staff replacement in the engineering sector, major foundries have closed shop, and presently, there is dearth of critical technology such as dye casting in Nigeria.

The firm stressed that the only thing special about the grinder is the engineering design. When a good design is done, Nigerian entrepreneurs often consider it better and cheaper to produce abroad in China/Japan if there is fund to order enough volume. This is unfortunate because attitudes of government and policies have not encouraged local production; and thus aggravating the tendency for good designs to be taken abroad for production while Nigeria serves only as a consumer market.

The firm views the decline in the engineering sector over the years (e.g. closure of foundries) as being due to lack of incentive for local production; lack of demand for local products; and generally poor government industrial policy. For example, the firm managers queried,

'why should government place import duty on sheet metal?... 'What else could sheet metal be used for other than industrial production?'

The firm claimed to have engaged in R&D with the Federal Institute of Industrial Research, Oshodi (FIIRO) and Raw Material Research and Development Council (RMRDC). It rightly observed that weak and small local firms cannot be involved in significant R&D. The firm also claimed to have worked with

Standard Organization of Nigeria (SON) to develop standards for yam pounding machine.

Power supply to the firm is about 50% from Power Holding Company and 50% self generated. Water is wholly self-supplied from a borehole. The firm considers reform in these sectors as yet unimpressive in improving infrastructure for industrial production.

Market for products

Capacity utilisation at the firm has been considerably low due to lack of market for products. The firm had 14 engineers in its employment between 1987 and 1993 when it was involved in a government supported project for the development of a Nigerian car. The project came to a halt when a change in the leadership at the Federal Ministry of Industry led to a change in government policy. The car project was terminated and project machines/equipment were moved into a public research institute established for the auto industry. The termination of the public-private partnership on the car project apparently led to non realisation of the dream of a Nigerian car.

The firm views market for product as the major problem of the engineering sector. It believes government policy should aim at stimulating market for the goods and services of the engineering sector. The firm claimed that a good attempt at this was made under the Directorate of Foods Roads and Rural Infrastructure (DFFRI) and the Petroleum Trust Fund (PTF) programmes. The DFFRI engaged the firm in production of road maintenance machines, water pumps for wells, tar boilers, and road rollers; while the PTF commissioned the firm to develop and produce corn shellers and rice hullers. The corn shellers and rice hullers were made from designs obtained from local research institutes.

Case Study 2: The cassava and food processing firm

Establishment and products

The firm is essentially a cassava processing company established in 1982 and refurbished in early 2005 after years of inactivity. The new company that emerged was incorporated in June 2006. Its products include:

- an intermediate product, cassava flour (for flour mills as input material to make 10% cassava-wheat flour). The flour can also be used as input material for glue making.
- consumer products such as semo-cassava (a bye product of cassava flour production); Fufu; Gari; Yam flour (seasonal); and Pure plantain flour (seasonal, from unripe plantain).

Yam and plantain flour production were still in experimental stages at the time of this study. Plantain flour is used for dietary purposes (e.g. for diabetic patients).

The employment structure of the firm is made up of about 50-60 casual workers (mainly women) who peel raw cassava and 40-60 monthly wage earners. On a regular basis the number of people in the firm's employment is about 100, about seven of which has a qualification of at least a Higher National Diploma (HND). The technical director has Masters Degree in food technology and the Managing Director is also a graduate.

Sourcing of input materials

The raw materials are completely locally sourced and are simply locally cultivated tubers. The firm went under in the period 1990-1991 due to lack of sustainable supply of cassava which is the main raw material. The liberalization policies that followed could not stimulate cassava production to meet the demands of cassava processing firms. The firm went as far as Abeokuta (a distance of about 200km) to procure cassava. This was too expensive for the firm and it eventually suspended production when the Abeokuta supply ceased to function due to the closure of TEXAGRIC in Ogun State. TEXAGRIC was a commercial farm, for which the firm was carrying out contract manufacturing, and it served as the supplier of the firm's major raw material, cassava.

The firm was revived by initiatives that involved partnership with the Ondo State Government in Southwest Nigeria. The firm is now jointly owned by government and Oke Ayo Farms Ltd (a private enterprise) with 49% and 51% equity respectively. The project is part of the cassava revolution policy of the Ondo State

Government. The state invested in the firm in February 2005 to the tune of N40million, and the Governor officially commissioned the firm in August 2005. The government fund enabled the firm refurbish its equipment and buy a new electricity transformer. Besides, the federal government policy of 10% cassava in wheat flour, tagged PICPE,[5]has encouraged more farmers to cultivate cassava. Ondo State Youth in Agriculture Programme (a poverty reduction scheme) has also significantly contributed to cassava production in the firm's locality. The state government allocated 500hectares of land to the firm for cassava plantation out of which the firm was cultivating 60hectares at the time of this study.

The firm reported that the small and medium enterprises equity investment scheme (SMEEIS), domiciled in the Central Bank of Nigeria and operated by commercial banks, has been inaccessible because the commercial banks are too selective and generally averse to investing in agro-allied projects. The banks are viewed as being generally sceptical about investments in farming and agro-processing. The firm believed that SMEEIS and other bank facilities that are meant for farmers still serve non-real farmers.

Manufacturing technology

Machines and equipment used by the firm can be completely fabricated locally. At the time of this study the firm employs a composite of imported and locally fabricated machines. The imported machines consist of peeling machine, rotary drier, and clarifier. The sifter, grinder and some of the presses were locally fabricated. The imported equipment were bought new in the early 1980s from United Kingdom. Prior to resumption of production in early 2005, the machines were all locally refurbished by Nigerian engineers and technicians. Maintenance of the plant was also carried out locally and the spare parts are available locally.

The firm's capacity utilisation was only 20% at the time of this study because the main equipment was originally designed for gari manufacturing. It was modified to produce cassava flour in response to government policy on 10% cassava flour in wheat flour. This was considered a significant impact of the PICPE policy.

Electric power supply remained a major problem: electricity available from the public electricity company for only about half of the production time. This was facilitated by the firm's own transformer supplied as part of government's equity. The firm runs on a heavy duty generator during power outages. The firm also has own borehole as source of water supply. These aggravate costs of production and are significant disincentives to production.

The firm partners with the food technology department of the Federal University of Technology Akure for quality assurance tests. There are also occasional links with the IITA in sourcing cassava stem cuttings and varieties, and visits to see new cassava processing technologies. The firm also indirectly benefited from cassava micro-processing programme organized by IITA and Niger Delta Development Commission (NDDC). The micro-processing project failed and the workers from the project brought the experience gained into this firm when employed by the firm.

Market for products

Market for the firm's product is vast in Nigeria. There are similar firms (e.g. two in the locality) in cassava processing but the market is far from being saturated due to Nigeria's large population. The firm has not embarked on export though it has been visited and encouraged by Nigerian Export promotion Council (NEPC). It intended to register with the NEPC in order to benefit from government's export promotion programmes. However, the firm claims that the export of cassava chips was unprofitable at the time of this study. For example, while 2.5tons of cassava is required to produce 1.0ton of chip; the raw cassava costs about N8000/ton and cassava chips sell for $120/ton on the international market.[6] Thus, apart from other production costs and freight charges, $189 worth of cassava is required to produce $120 worth of chip! Meanwhile the firm plans to export plantain flour, yam flour and fufu in future.

Access road to the plant was graded by the local government. The state government provides free publicity of the firm's activities at various state government fora. State government involves the firm in its industrial promotional activities as an exemplary case of public-private partnership in industrial production. Environmental pollution is not a serious concern at the plant. The wastes are biodegradable and released into the uncultivated land around the plant. However, most of the wastes are either sold or picked up free by people that use them for animal feeds or as compost to grow edible mushrooms.

The firm claimed that government does not take stakeholders along in making policy decisions. It thus felt scared that the PICPE policy may not endure if it is not backed up by law, and view this as a significant threat to the future of the firm. The firm also stressed that establishment of cassava plantation should be encouraged by government. Policy should aim at reducing costs of cassava farming so that cassava can be obtained at much lower costs than it was at the time of this study. For example, diesel that tractors (and industry) use should be cheaper than petrol, land could be cleared free for farmers, and subsidised farm inputs could be made available at the right time to farmers. The stringent conditions for loans to real farmers could also be relaxed.

Case Study 3: The flour mill

Establishment and products

The firm was incorporated in October 1979 but started operation three years later in October 1982. The total staff strength at the time of this study was about 150 including four expatriates, and four engineers/chemists. The firm has a standard laboratory for quality assurance.

The firm is 80% Nigerian owned while foreign equity is 20%. The foreign partners are from the Switzerland. The major product of the firm is flour which is largely an intermediate product for firms in the bakery and confectionery subsector. Other products include semolina and wheat offal.

Sourcing of input materials

The major economic policy that has affected the firm in recent times is the policy on 10% cassava in wheat flour under PICPE.

Source of cassava is local firms while wheat is imported. The firm obtains cassava usually by advertising the required quality of cassava flour. Local firms producing cassava flour responds, and orders delivered by these firms are subjected to laboratory tests to ensure that the cassava flour meets the specifications for blending with wheat flour.

Manufacturing technology

The milling machines are all imported with the assistance of the foreign partners. Spare parts are also imported. The firm has an in-house repair workshop while major repairs are carried out by the foreign technical partners. On a few occasion the MTNL Osogbo played significant roles in carrying out minor repairs and fabrication of spare parts. The plant engineers have attended courses and workshops at the IITA especially in connection the 10% cassava in flour policy. The firm's products are NAFDAC and SON compliant and arrangement are on to make the products attain ISO 9000 certification. As at the time of this study, the firm runs three shifts per day. Power supply from the national grid is available for only one out of the three shifts.

Road to the plant site was constructed by the firm in 1982, and the firm has maintained the road until about five years ago when the government took over its maintenance. Water supply to the firm is by own borehole facility. The production system is a dry process and there are no flour dust leakages or effluents.

Market for products

Production output is sometimes constrained by sales. The shelf life of flour is relatively short largely due to insect infestation of flour. This necessitates keeping inventory of unsold stocks to the barest minimum. Thus, if sales decline production is put on hold until the market improves.

There are about 16 active flour plants in Nigeria and competition is relatively stiff. Competition is very sensitive to quality of the products. Two years ago the quality of the firm's product was discovered to be less than those of its major competitors. This significantly affected the firm's sales turnover. The problem was traced to the low quality wheat imported by the foreign partners.

The wheat and cassava flour inputs were subsequently routinely subjected to rigorous laboratory tests. This enabled the solution to the problem and demand for the firm's products has since increased. Capacity utilisation at the plant was less than 60% in 2004, increased to 70% in 2005, and attained 75% in 2006. The improvement in capacity utilisation was achieved through reorganisation of the firm and improvement in the quality of the input materials. The market for the products is only local and there is no evidence that the firm has export plans. There is abundance of local market for products.

The main challenge identified by the firm is how to reduce high cost of production arising mainly from use of diesel generators for two out of three shifts per day; and relatively high overhead costs.

Policy mechanisms and innovation capacity of SMEs in the agro-industry

From the foregoing, the case studies of SMEs are varied in their illustration of the state of local technological capability in Nigeria's agro-industry and the role of policy as an intervention mechanism for strengthening the innovation capacity of SMEs in the agro-industry. The case study on the engineering sector revealed that impact of Nigeria's economic policy on the engineering sector might be very marginal. The case study suggests that the sector has since been declining before the advent of recent economic policy reforms and the end of the decline is not yet in sight. In particular, sourcing of input materials, which are mainly imported, and the stimulation of market for products are the main challenges that should be addressed to stop the decline and revive the sector.

The two cases of SMEs from the agro industrial subsector demonstrated a significant impact of policy in promoting the emergence of innovative agro-industry. Case II showed that public-private partnership could be beneficial in stimulating SME development, technological upgrading, and employment generation. In this respect, the case study appeared to be an experiment that would be proven at least in the medium term when the firm is hopefully able to provide good returns on investments. The policy impact so far in case II has led to significant technological upgrading entailing both process and product innovations. It is also important to stress that these good outcomes essentially depend on one critical aspect of policy reform, viz., 10% cassava in wheat flour under PICPE. The sustainability of this policy is a very crucial factor to this case study. Similarly, case III has also benefited significantly from the cassava policy under PICPE. However, there has been no demonstrable technological impact in terms of improved process or significantly new product. The only impact is the opportunity to improve skills or acquire skills needed to successfully achieve the objective of 10% cassava in flour. In this respect, it was reported that the plant engineers have attended courses and workshops at the IITA.

Common to all the three case studies are the need to provide infrastructure support to the firms. This is a confirmation of the widely acknowledged problem of poor infrastructure as major constraint on industrial production in Nigeria. The case studies demonstrated that the impact of policy reform, especially in the electric power sector is yet to significantly affect the manufacturing sector. It is imperative that future policy reform measures should aim at accelerating the pace of infrastructure development especially in the areas of electric power, transportation and water supply to industrial plant sites.

6. Concluding Remarks

In this chapter, we have attempted an analysis of the links between local technological capability, innovation capacity and their imperative for the development of the agro-industry in the quest for structural transformation of the Nigerian economy. In this concluding section, we extend the discourse by presenting the prospects for improving the technological base and innovation capacity of the Nigerian agro-industry.

As observed by AfDB (2000), the transformation of African agriculture requires a shift from the highly diversified, subsistence-oriented farming activity towards a more commercially-oriented agriculture with improved access to markets and agro-industry. AfDB rightly stresses that the transformation would involve a greater reliance on input and output markets and increased integration of agriculture with other sectors of the domestic and international economies. In addition, it entails a more efficient and balanced use of indigenous knowledge and "modern" scientific knowledge. For Nigeria, the prospects of achieving the required structural transformation that would result in increasing the technological base, innovation capacity and human capacities for agro-industrial development may be highlighted from the insights gained from the discourse in this chapter.

Poverty reduction remains a major goal of economic policy in Nigeria. Building a dynamic agro-industry that is innovation driven is a potent instrument for achieving pro-poor growth. Generally speaking, the agro-industry is part of the industrial innovation system, which is notably weak and unequipped to respond to the technological challenges facing the Nigerian industrial sector. This is acutely pronounced in the SMEs and has been illustrated by case studies of the state of technological capability in the agro-industrial subsector. The technological trajectory in industrial innovation is still much at the initial conditions of articulating the frameworks for technological learning, building infrastructure for R&D, and providing adequate and appropriate incentives for industrial innovation. Consequently, the existence of technological innovation in the Nigerian agro-industry has been limited to anecdotal incidence of industrial innovation.

From the analysis in this chapter, specific suggestions for increasing technological base and improving innovation capacity in the Nigerian agro-industry would include:

- The Nigerian industrial and technology policies should be reviewed with active participation of stakeholders with the objective of making them innovation focused. The review should principally aim at integrating industrial and technology policies in an innovation policy that details specific sectoral strategies for stimulating a dynamic industrial innovation system with a focus on agro-industry as a major component.
- Strong networking and interactions among actors (industrial stakeholders) should be encouraged and promoted. Expenditures on networking activities especially involving SMEs

should be tax-deductible.

- R&D institutions should be rehabilitated and provided with infrastructure that will make them attractive for skilled experts at home and in the Diaspora. A major boost in R&D spending is required, and the managers of R&D institutions should be re-orientated to adopt a system view of innovation capacity development as the main rationale for R&D rather than research just for the sake of adding to knowledge and publishing. This will facilitate joint initiation and execution of R&D with firms, which should be provided with generous incentives for private sector R&D, especially to address local needs. This will also greatly enhance the development of human capacity required for R&D in the public and private sectors with good prospects for significant improvement in public-private sector interactions on demand-driven R&D projects that address societal needs.
- There should be increased investment in knowledge and skills development by public and private sectors.
- Strong export-orientation of firms is crucial for competitiveness especially on the medium to long term. From the discourse In this chapter, industrial policy should be specifically targeting the promotion of innovative and export-oriented SMEs in the agro-industrial subsector.

Notes

1 See Radosevic (2009) for a recent review of the SAP policy and how it relates to the post-Washington consensus approach to development of technological capability in developing countries.

2 Meeting of OAU/STRC Inter-African Committee on Science and Technology for Development, held at International Institute of Tropical Agriculture (IITA), Ibadan, Nigeria, 26-30 August 1996.

3 In a study of five Sub-Saharan African (SSA) countries, it was found that 43% of the increase in employment during the 1980-91 period occurred in small enterprises employing less than 50 people. While in Colombia, India, Indonesia, Kenya, Philippines, Tanzania and Zambia, SMIs with up to 50 people employ over half the industrial workforce (Riddell, 1990).

4 These case studies were carried out in the third quarter of 2006 as part of a survey of SMEs in Nigeria by the Nigerian Institute of Social and Economic Research (NISER).

5 PICPE is the Presidential Initiative on Cassava Production and Export which was officially launched in 2004 to promote the diversification of the economy through industrial processing of cassava to add value and achieve significant export of cassava products (CBN, 2007; Adeniji, 2007).

6 Exchange rate was US$=N127 at the time of this study in 2006.

References

Adeniji, A. A. (2007), 'Cassava Varieties for Ethanol,' *Paper presented at the first International Ethanol and Biofuels Conference*, Abuja, Nigeria, 29-30 March.

Adeoti, J. O. (2002a), 'Building Technological Capability in the Less Developed Countries: The Role of a National System of Innovation', in *Science and Public Policy*, 29(2): 95 -104.

Adeoti, J. O. (2002b), *Technology and the Environment in Sub-Saharan Africa: Emerging Trends in the Nigerian Manufacturing Industry*, Aldershot, UK: Ashgate Publishing Ltd.

Adeoti, J. O. and Olubamiwa, O. (2009), 'Towards an Innovation System in the Traditional Sector: The Case of the Nigerian Cocoa Industry,' in *Science and Public Policy*, 36(1): 15-31.

Adeoti, J. O., Odekunle, S. O. and Adeyinka, F. M. (2010), *Tackling Innovation Deficit: An Analysis of University-Firm Interaction in Nigeria*, Ibadan, Nigeria: Evergreen Publishers.

AfDB (2000), *Agriculture and Rural Development Sector* Bank Group Policy, Tunis, Tunisia: African Development Bank.

CBN (2007), *Central Bank of Nigeria Annual Report and Statement of Accounts 2006*, Abuja: Central Bank of Nigeria.

Clark, J., McLoughlin, I., Rose, H. and King, R. (1988), *The Process of Technological Change: New Technology and Social Choice in the Workplace*, Cambridge: Cambridge University Press.

Ekundare, O. (1973), *An Economic History of Nigeria 1860-1960*, London: Methuen & Co. Ltd.

Freeman, C. (1995), The 'National System of Innovation' in Historical Perspective,' in *Cambridge Journal of Economics*,19(1): 5-24.

Freeman, C. and Soete, L. (1997), *Economics of Industrial Innova-*

tion, London: Pinter Publishers.

Goel, V. K., Koryukin, E., Bhatia, M. and Agarwal, P. (2004), 'Innovation Systems:World Bank Support of Science and Technology Development', *World Bank Working Paper No.32*, Washington D.C.

Jalilian, H., Tribe, M. and Weiss, J. (eds.) (2000), *Industrial Development and Policy in Africa: Issues of De-industrialization and Development Strategy*, Cheltenham, UK: Edward Elgar.

Kim, L. (1997), *Imitation to Innovation -The Dynamics of Korea's Technological Learning*; Boston, Massachusetts: Harvard Business School Press.

Lall, S. (1987), *Learning to Industrialise -The Acquisition of Technological Capability by India*, London: Macmillan Press.

Maduemezia, A (1996), 'Development of Indigenous High Technology in Africa', *Paper presented at the OAU/STRC Expert Meeting on the Promotion of Indigenous Technologies in Africa*, 26-30 August, IITA, Ibadan, Nigeria.

Muchie, M. and Baskaran, A. (2009), 'The National Technology System Framework: Sanjaya Lall's Contribution to Appreciative Theory,' in *International Journal of Institutions and Economies*, 1(1): 134-155.

Nelson, R.R. and Winter, S. G. (1982), *An Evolutionary Theory of Economic Change*, Cambridge, MA: The Belknap Press of Harvard University Press.

Ogun, O. (1995), 'Country Studies: Nigeria,' in Wangwe, S.M. (ed.), *Exporting Africa: Technology, Trade and Industrialisation in Sub-Saharan Africa*, UNU/INTECH Studies in New Technology and Development, London: Routledge.

Okejiri, E. (2000), 'Foreign Technology and Development of Indigenous Technological Capabilities in Nigerian Manufacturing Industry,' in *Technology in Society*, 22(2): 189-199.

Okigbo, B. N. (1996), 'Promotion of Indigenous Technologies in Sustainable Development in Africa', *Keynote address delivered at the OAU/STRC Expert Meeting on the Promotion of Indigenous*

Technologies in Africa, 26-30 August, IITA, Ibadan, Nigeria.

Riddell, R. C. (ed.), (1990), *Manufacturing Africa: Performance and Prospects in Seven countries in Sub-Sahara Africa*,Portsmouth, UK: Heinemann.

Stewart, F. (1983), 'Facilitating Indigenous Technical Change in Third World Countries,' in Fransman, M. and King, K. (eds.), *Technological Capability in the Third World*, London: Macmillan.

Radosevic, S. (2009), 'Policies for Promoting Technological Catch-up: Towards a Post-Washington Approach,' in *International Journal of Institutions and Economics*, 1(1): 23-52.

Chapter 4

Determinants of Technology Adoption in Ugandan Agricultural Manufacturing Firms

Niringiye Aggrey

1. Introduction

This study examined the relationship between the decision to adopt new technology and its determinants. Total factor productivity (TFP) differences can explain cross-country variation in real incomes to a large extent, and that technology, defined broadly as knowledge relevant to production, is an important determinant of TFP across countries and firms (Easterly and Levine, 2001 and Prescott, 1998). There is growing evidence suggesting that knowledge, innovation and ongoing technological change are strong determinants of productivity, growth differentials, as well as the ability of countries to benefit from globalization (Oliner and Sichel, 2000; Timmer and van Ark, 2005). Why do some firms adopt new technologies faster or at higher rates than others? Knowing the determinants of adoption can help predict adoption patterns, for example, which firms are likely to be first movers, and therefore leap ahead when new technology is introduced or this may help us understand how to encourage adoption of beneficial technologies?

The empirical analysis of technological diffusion has concentrated on general tests concerning the impact of different explanatory variables on the adoption decision of firms. The debate has emphasized the role of market structure and firm size, following the seminal analysis by Schumpeter (1942). However, most studies on technology adoption have been conducted in developed countries like the US and Western Europe. Although there have been a number of international studies that have identified various country specific technology adoption issues, there is a dearth of knowledge regarding factors that influence technology adoption in developing countries. It is important for countries to be studied individually so that technology adoption practices that exist can be known and understood. The focus of our study is to extend existing models used in developed countries and see whether the determinants of technology adoption are similar in developing countries. There are significant differences between organizations in developed and developing countries. Organizations in developed countries have access to more resources and face more competition than organizations in developing countries. In addition, organizations in developed countries have greater access to technology than those in developing countries. Since there are inherent differences between organizations in developed and developing countries, we consider it important to examine the applicability of models developed for technology adoption in developing countries.

This analysis is important as it contributes to the empirical and theoretical debate on the determinants of technological adoption and considers a more detailed set of explanatory variables which enable us to set up a well-specified empirical model. Uncovering the factors driving technology adoption and diffusion is important and relevant for both research and policy. First, in contrast with well established theoretical literatures on new technology adoption, firm-level empirical evidence on adoption of new technology in agricultural manufacturing firms is still limited. Second, from the policy perspective, to the extent that a wide and fast adoption of technology is desirable, it is essential to understand what factors are likely to increase the adoption of technology.

Empirical evidence on the relative impact of various determinants of technology adoption in developing countries at the micro level is still scarce, due in part to the limited availability of data. This study aims to contribute to filling this gap by investigating how the "investment climate", namely the incentive regime faced by a firm (related to competition and ownership), access to complementary inputs (skills, credit, and infrastructure), and access to international knowledge affect firms' technology adoption. Our analysis exploits a unique dataset – investment climate survey of firms in Uganda by World Bank in 2003, which combines information on the investment climate with measures of technological update at the firm level.

Specifically, the question we investigated In this chapter is: what factors affect the chances of technology adoption at the agricultural manufacturing firm level? This paper contributes to the empirical literature on determinants of new technology adoption. Specifically, in comparison to existing cross-section studies, we estimated an improved econometric model by using investment climate survey panel data and account for firm heterogeneity and time specific effects which reduce the omitted variables bias. Our results showed that access to complementary inputs; skilled labour, location in the main city, being young, not being a member of a trade union - are positively associated with technology adoption. Results for the relationship between market incentives and technology adoption show that foreign ownership and competition from imports increases the likelihood of adopting technology. Our results also indicated that access to international knowledge through participation in export market is positively correlated with technology adoption, which can suggest the relevance of international technology spillovers.

The remainder of this chapter is structured as follows. In Section 2 we discussed the conceptual framework. Section 3 outlined the empirical strategy and model specifications. Section 4 presented and discussed the main results and section 5 summarized the main conclusions.

2. Conceptual Framework

It is assumed that firms make a profit-maximizing cost-benefit assessment of different alternatives when deciding whether to adopt new technology. A firm decides to adopt a particular new technology if the corresponding expected net payoffs are larger than those of the alternatives, including the alternative of not adopting it. Besides the cost of the technology itself, different investment climate factors may affect the expected net payoffs from the adoption of modern technology. This study draws on theoretical and empirical literature to lay out three groups of investment climate factors, including, access to complimentary inputs, market incentives and access to international knowledge that may affect those payoffs. A set of hypotheses is also implicitly defined, that will subsequently be tested using firm-level investment climate survey data from Uganda.

A firm's access to complementary inputs can affect both the adjustment costs as well as the benefits that the firm derives from the adoption of new technology. First, both theoretical models and empirical evidence document how low labour skills can delay a firm's modern technology adoption due to the inability to operate advanced equipment - which tends to be skill-biased - or because learning is technology-specific and there is a high cost of retraining workers (Alesina and Zeira, 2006). Following the seminal paper by Nelson and Phelps (1966), a large empirical literature has focused on the relationship between human capital and new technology adoption. Firm-level evidence suggests that firms using advanced technology require high skilled workers (Doms et al., 1997). The lack of highly qualified managerial capacity (or entrepreneurship) can also constrain the firm's adoption of advanced technology as it reduces the information on available technological solutions for its needs and could increase its adjustment costs (Cohen and Levinthal, 1989). It is also argued that, even though capital goods are considered good collateral in developed countries, failure to regulate the movable collateral regime in developing countries may constrain smaller firms' access to credit, restricting their adoption of new technologies. Depending on the technological gap between existing and new technologies and on the associated adjustment costs, complementary physical investments may be needed, in which case access to credit becomes an even more important input for technological update. Based on the potential impact of transport infrastructure in the production process, infrastructure availability and quality can also play a decisive role in a firm's decision to acquire and use new vintages of capital. The importance of physical infrastructure for firm productivity and economic growth finds strong support from empirical work for OECD countries as well as developing countries (Roller and Waverman, 2001; Esfahani and Ramirez, 2003).

Drawing on this literature, a recent study by the World Bank (2008) concludes that the diffusion of international knowledge and technology to a developing country depends on two main factors: (i) the extent of its access to international knowledge and technology (via trade, FDI, contacts with its Diaspora, and other communications channels) and (ii) its ability to absorb and adapt international knowledge and technology. This absorptive capacity is determined by the technological literacy and skills of the country's labour force, the availability of financing for innovative firms, sound governance, and a good business environment. Moreover, given the various market imperfections associated with technology and knowledge diffusion, including increasing returns to scale and technological spillovers, that study also finds scope for carefully targeted proactive policies aimed specifically and enhancing technology transfer.

The firm's technology adoption decision is also likely to be determined by market incentives which refer to four core aspects of a competitive market economy. First, private ownership and control establish profit-maximization as the 'modus operandi' of the firm and should drive investment in the most productive technologies with ensuing productivity gains (Brown et al., 2006). Second, an important additional condition is strong product market competition from domestic and foreign rivals and/or demand from consumers which encourage incumbents to direct their investments towards new and more productive technologies - product upgrades and/or costs reductions- rather than towards rent-seeking activities (Baumol, 1990; Aghion and Schankerman, 2004).Third, sound legal institutions that guarantee adequate protection of property rights and contract enforcement are critical for growth and development of market economies (Djankov et al., 2006). Acemoglu et al. (2007) showed how greater contractual incompleteness leads to the adoption of less advanced technologies and how that incompleteness may generate sizable productivity differences across countries with different contracting institutions. Finally, firms need to be confident that they will be able to reap the benefits from their technology adoption investments, as in the initial stages of the adoption process investments in new technology may even generate productivity losses related to capital and skills specificities, adjustment costs, and learning-by-doing (Pavlova, 2001). Adjustment and learning costs are proportional to the technology gap. In this sense, a predictable economic and regulatory policy environment is a fourth key market incentive affecting the firm's decision to adopt new technology.

Competitive pressure has been identified as an incentive to innovate and adopt new technology (Porter, 1990). Firms facing stronger competition are more inclined to innovate and adopt new technologies, in order to strengthen their performance and survival rate. Several studies have shown that competitive pressure is positively associated with technology adoption (Dasgupta et al., 1999; Hollenstein, 2004). In contrast, other papers find no significant effect of competitive pressure on technology adoption (Lee, 2004).

Another stylized fact supported by a large empirical literature is that larger firms are more likely to adopt new technologies faster. Firm size is commonly used in the empirical literature on new technology adoption because it is easy to observe and it serves as a proxy for several things (Geroski, 2000). Large firms can earn higher profits from adopting new technology in comparison to small firms. Given the risks and costs of early adoption, they are in a better position to adopt new technology because they have fewer financial constraints and are likely to be less risk averse. They might be more motivated and able to innovate in order to pre-empt smaller rivals. Also, the scope for innovation complementarities is likely to be greater in larger firms. A positive correlation between firm size and ICT adoption is found in a number of empirical studies (Fabiani et al., 2005). Other studies, in contrast, have found a weak or not significant relationship between firm size and the adoption of ICT (Lefebvre et al., 2005; Love et al., 2005). Furthermore, Hollenstein (2004) showed that this relationship might be non-linear.

The firm's technology adoption decision relates to its access to international knowledge which can be transferred via FDI or through the participation in export markets. Openness to trade and FDI have been shown to be critical mechanisms for knowledge diffusion across countries (Coe et al., 1997; Keller, 2004). Openness to FDI introduces competitive pressure on firms but also allows firms to be exposed to global best practice technology and management techniques (Djankov and Hoekman, 2000; Arnold and Javorcik, 2005). Domestic firms that participate in export markets may increase their knowledge absorption from the interaction with buyers in advanced markets (Wagner, 2007). Thus, firms with foreign ownership and those participating in export markets are likely to face lower costs and be more likely to engage in the adoption of advanced technology. It has been argued that firms exposed to international competition in export markets are more inclined to innovate and adopt new technologies.

Hall and Khan (2003) summarized the literature on the determinants of technology adoption mainly in developed economies by emphasizing the importance of microeconomic factors such as labour skills, access to capital goods, network effects, market structure, firm size, regulatory environment in explaining broad patterns of technology diffusion. When examining cross-country differences, the authors note that other macroeconomic factors, such as the level of economic development, market size (Helpman and Krugman, 1986), openness to trade and FDI (Keller, 2002), and culture can help explain differences in the level of technology diffusion across countries. Also, Comin and Hobijn (2004) showed that the most important determinants of the speed at which a country adopts a broad range of technologies are human capital endowment, government type, degree of openness to trade, and adoption of predecessor technologies.

3. Methodology

The empirical strategy consists of testing the various hypotheses implicitly discussed in the conceptual framework, namely, that firms face lower costs of adjustment and higher benefits from technology adoption and thus are more likely to engage in the adoption of advanced technology if they have better access to complementary inputs, face stronger market incentives, and have better access to international knowledge. It is assumed that profit-maximizing firms decide whether or not to adopt advanced technology if the benefits from this decision are larger than its costs. The adoption of technology, which depends on unobserved net benefits in firm i, at time t, is modelled as a function of access to complimentary inputs, market incentives and access to international markets in a linear form. The basic model specification is as follows:

$$INVESTECH_{it} = \beta_0 + \beta_1 COMPINPUT_{it-1} + \beta_2 INCENT_{it-1} + \beta_3 IK_{it-1} + v_i + \mu_{it} \quad (1)$$

where $INVESTECH_{it}$ is a dummy variable taking the value of 0 if firm i did not invest in technology and 1 otherwise, β_0 is a constant, β_1, β_2, and β_3 are vectors of coefficients of explanatory variables for complimentary inputs ($COMPINPUT_{it-1}$), market incentives ($INCENT_{it-1}$), and access to international knowledge (IK_{it-1}) respectively and v_i and μ_{it} are firm fixed effects and the error term respectively. To proxy for complementary inputs, we considered measures of human capital, access to finance, and the quality of infrastructure. To proxy for market incentives, we considered, firm age, firm size, competition from imports, labour regulation, union membership and monopoly power. To proxy for access to international knowledge, we considered foreign ownership and firm participation in export markets.

Discussion of variables

Our first dependent variable was a dummy, indicating whether or not a firm invests in technology, lending itself to a probit analysis. The adoption of advanced technology is measured by investment in technology as reported by firms during the investment climate survey. Firm size, was measured as the number of employees in each firm. The expected effect of this variable was positive following theoretical arguments and previous empirical studies. This expectation was derived from the assumption that new technologies may have positive scale effects, implying that adoption is more profitable for larger firms and thus the probability of adoption increases as firm size increases. In addition, size may operate through finance availability, in that small firms are more likely to be credit rationed than larger firms in financial markets. Access to finance dummy variable was equal to one if the firm had a bank loan or overdraft and zero otherwise.

As a proxy for human capital we used the proportion of skilled labour force to total workforce and the education of the top manager. Firm's labour force skills variable was a proxy of the firm's human capital level. The expected effect was positive as new technologies can more easily be introduced in firms with a more qualified labour force. Corruption may act as a barrier to the adoption of new technology may take different forms, such as, bribes that must be paid, but essentially represent a distortion in the relative payoffs facing the entrepreneurs in favour of the sub-optimal technology (Parente and Prescott, 1994). The education of the manager was believed to be positively related to investment in technology, implying that better educated managers do have greater possibilities of borrowing to finance new technologies. Better educated managers would find it easier to present a plausible case for a loan to an outside body (Abor, 2008). Overall, the level of education appears to have an important positive impact on micro and small enterprises' debt-raising capacities (Green et al., 2002). Level of education of highest manager was constructed as a binary with dummy equal to one if manager completed education above secondary education and zero otherwise. We expected to find a positive coefficient and significantly different from zero, between human capital indicators and probability of investing in technology.

Two proxies for the quality of infrastructure were used, that is power outages and perception about the adequacy for infrastructure. Public sector investment (in roads, schools, hospitals, public utilities and other business support services) facilitates a better business environment. More public sector investment in infrastructure would therefore attract more private investment. In addition, manufacturing firms close to the capital city would have easier access to debt finance than those located outside the capital city. We controlled for location effects, including a dummy for the firms located in the main city and zero otherwise and expected a positive relationship between location and technology adoption. Power outage was measured by the number of power outages or surges from national grid suffered. Inadequate infrastructure was measured by dummy equal to one if the response given by a firm on whether inadequate infrastructure is among the three major constraints of doing business in Uganda and zero otherwise. Labour regulation and union membership were measured using dummies.

In addition, basic economic theory informs us that monopolies will prefer to restrict output in order to earn supernormal profits and they are therefore unlikely to invest in new technology. We used a dummy variable indicating whether any of three main product lines have more than 60% of the market share. Furthermore, to proxy for foreign competition, a dummy equal to one if firm indicated that it faces "intense competition" from foreign firms; either through imports or FDI, was used.

Output growth rate captures the effect of output growth on the probability of adoption. The expected sign on the coefficient of this variable was positive as an expanding (or at least steady) output allows the allocation of more resources to technology adoption. Firm age may affect access to finance and thus be associated with firm-specific capital costs. Longer-established firms could arguably be thought to have access to lower-cost capital and greater experience. If age is proxying cost, it should affect both the decision to invest and the amount of investment undertaken. Age was estimated by the year of the survey minus the year when the firm started operations. The inclusion of output growth and firm age variables is based on the prior assumption that fast-growing firms, new firms or firms with outdated equipment are more likely to adopt new technologies or use them more intensively (Karshenas and Stoneman, 1993). While we expected to find positive relationship between output growth and technology adoption, the relationship between firm age and technology adoption is ambiguous. This follows from the fact that while older firms may show higher rates of technology adoption and use due to learning effects, younger firms may be better placed to adopt recent available technologies (Barbosa and Faria, 2008).

Export participation and foreign ownership are included as measures of access to international knowledge. Firms participating in export business tend to be more diversified and as such are capable of accommodating more debt capital (Abor, 2008). Thus, as firms engage more in exporting activities, they tend to employ more debt and therefore likely to invest more in technology. Export status was constructed as a binary that equal to one if a manufacturing firm was engaged in exports and zero otherwise. Foreign ownership was captured by a dummy of one for a firm whose foreign ownership was over 50%. We expected a positive relationship between export participation, foreign ownership, and the probability of investing in technology.

Econometric issues such as endogeneity, and measurement errors, were addressed. The question of the causality, for example, from firm size to investment in technology was addressed. Are firms, for example, investing because they are large, or are investing firms in technology becoming larger? Probably both explanations hold to some degree, and may be able to explain why some firms are unwilling to invest in technology. The same applies to other endogenous variables. The endogeneity issues were addressed by using lags of variables that are endogenous to reflect the manager's information set at time t. On measurement errors, observations with extreme values were deleted. Ideally we would also have liked to control for firm fixed effects, however the panel data with a short time dimension does not make this an appealing option.

Data

The data was obtained from survey data that were collected from interviews during 2002- 2003, as a part of the World Bank Regional Program on Enterprise Development (RPED), in collaboration with the Uganda Manufacturers' Association Consulting Services (UMACIS). The sample frame consisted of all manufacturing firms with ten or more employees in the 2001/02 Census of Business Establishments. A stratified random sample was drawn from this frame with location, size and sector as the stratification dimensions. The stratification yielded 56 populated clusters. The sampled firms represented a sampling rate of 41% and account for about 70% of employment of the sampling frame. The data is reliable and of reasonably good quality.

In addition to information on investing in technology and investment in new machinery and equipment, the survey collected information on perceptions on firm ownership structure, skill of workers, participation in export markets, and access to finance, corruption, and firm size among others. The relevant sample included 122 agricultural manufacturing firms that had complete data on most variables of our interest. Although the data is not strictly comparable to surveys in other countries, useful comparisons were made between the results obtained from the survey data and those obtained in other African countries.

4. Results and Discussion

In this section, we discussed our main results obtained from probit estimation of Equation (1) for investing in technology. The main specifications is estimated based on the investment climate survey panel sample which allows us to control for unobserved firm invariant characteristics which may influence both the technology adoption decisions as well as their correlates. The probit regressions with random effects allowed us to analyze how changes in investment in new technology by agricultural manufacturing firms in Uganda are related to changes in complementary inputs, market incentives, and access to international knowledge. The disadvantage of this approach is the short panel dimension of the data (only two years of data per firm) and the smaller sample size of 122 agricultural manufacturing firms. With these caveats in mind we discussed the regression results presented in Table 1.

Table 1: Probit Regression Estimates: Dependent variable - Investment in Technology

Equation	1	2
Variable	Coefficients(z-values)	Coefficients(z-values)
Constant	-8.091(-7.04)***	-6.132(-2.64)***
City dummy	2.126(2.94)***	3.178(2.78)***
Export dummy	1.433(1.63)*	-1.114(-.87)
Ln(size)	3.606((6.08)***	5.752(3.91)***
Ln(age)	-1.761(-2.65)***	-5.449(-3.02)***
Education manager	-5.290(-.81)	-2.620(-2.11)
Foreign ownership	1.599(2.38)**	5.746(3.08)***
Finance access	-.151(-.24)	.573(.51)
Unionised	-.0205(-1.08)	-.0182(.48)
Skilled proportion	.0451(4.51)***	.0783((3.83)***
Sales growth	.0153(2.21)**	-.000948(-.05)
Competition imports	4.503(3.46)***	2.482(.71)
Power outage		-.355(-.37)
Monopoly		.428(.35)
Infrastructure problem		-3.698(-2.72)***
Labour regulation		-0.0241(-.8)
log likelihood	-63.948	-29.332
Wald chi2	86.70	41.96
Prob>chi2	.000	.000
No. of observations	223	112

***, **, and *: indicates significance at 1%, 5% and 10% respectively.

The regression results showed that larger manufacturing firms registered a higher conditional probability of investing in new technology, all else equal. This finding is consistent with theory and earlier empirical findings by Fabiani et al. (2005). It was expected that the advantages of large-scale production would positively influence technology adoption and vice versa, so that smaller firms would find it difficult to adopt technology than large ones. A large firm in terms of establishment and output levels is definitely better placed to acquire and adopt new technologies than a smaller firm. These findings on size are consistent with two related arguments. First, large firms may benefit from economies of scale in the adoption of new technology and have more capacity to finance technology adoption in imperfect financial markets settings (Cohen and Klepper, 1996). Second, the economies of scale from which large firms benefit allow them to operate with a more efficient division of labour, resulting in better conditions for mechanization and technological upgrade.

The proportion of skilled workers variable is significant and with the expected sign. This evidence suggests that the skill of the labour force acts as a requisite for investment in technology. This evidence is consistent with earlier findings obtained by Chun (2003) who showed that highly educated workers are more likely to implement new technologies. Bartel and Sicherman (1998) also find that industries with higher rates of technological change require highly skilled workers. Our findings provide indirect evidence of skill-biased technological change by showing the importance of complementary investments in skills for technology adoption in Ugandan agricultural manufacturing firms. In sum, firms with higher shares of professionals have better capacity to process information, i.e., to identify technological problems and solutions.

By the results, age is also shown to be consistently negatively associated with the probability of investment in technology. This finding is consistent with the argument by Barbosa and Faria (2008) who argued that while older firms may show higher rates of technology adoption and use due to learning effects, younger firms may be better placed to adopt recent available technologies.

The result also showed that foreign ownership of firms is consistently significant and with the expected sign. It has been argued that foreign owned firms are more likely to be early adopters of new technology and they are important channels of new technology diffusion (Barbosa and Faria, 2008). Export participation, which was also used as a proxy for access to international knowledge, was also shown to be positively associated with the probability of investing in technology, although the relationship is weak.

As expected, sales growth was shown to be positively associated with the propensity to invest in technology. This result may suggest that as output grows, a proportion of that increment is allocated to investment in technology. Competition from imports is shown to be positively associated with the propensity to invest in technology. Firms that reported that they face intense competition from imports as one of the major obstacles of doing business showed a positive association with propensity to invest in technology. This might suggest that, to compete in the newly deregulated market, firms facing intense competition were more likely to invest in future production capacity by investing in technology.

Inadequate infrastructure problem is significantly negatively associated with the probability of investment in technology. The results also showed that firms located in the main city were more likely to invest in technology, thus suggesting that infrastructure is a major determinant of investment in technology.

Variables, such as monopoly power, labour regulation, power outage, and union membership had the expected signs but the coefficients are not significantly different from zero. Managerial education, and financial access are negatively associated with investment in technology but the association is not significant.

5. Conclusion

This study contributes to the empirical analysis of determinants of new technology adoption. The adoption of new technologies in Ugandan agricultural manufacturing firms was analyzed using data on 122 firms involving investment in new technology and purchase of new machinery and equipment in the period 2000–2002. In comparison to existing cross-section studies, an improved econometric model was estimated using firm-level panel data from the Ugandan agricultural manufacturing firms. Investment climate factors were divided into three groups – complementary inputs, market incentives, and access to international knowledge - and estimated several strong correlations of those factors with the adoption of advanced technology as proxied by investment in technology.

This study found that firms with access to the appropriate complementary inputs(i.e., skilled labour, good infrastructure), market incentives (i.e., large firms, young firms), facing competition from imports, experiencing output growth and have access to international knowledge through foreign ownership and export participation, are more likely to invest in technology. Our main findings can be summarized as follows: the propensity to invest in new technology is higher for larger firms, younger firms, fast-growing firms, skill-intensive, firms participating in export markets, foreign owned firms, and firms with access to good infrastructure.

Our findings have implications for policy reforms aimed at enabling faster and greater adoption of advanced technologies by the private sector in Uganda. To increase technology adoption by agricultural manufacturing firms in Uganda, there is a need for complementary reforms of the investment climate: improvements in the access to international knowledge and an increase in the availability of inputs complementary to knowledge within a wider framework of improving market incentives.

References

Abor, J. (2008), 'Determinants of the Capital Structure of Ghanaian Firms', *African Economic Research Consortium Research PaperNo.176*, Nairobi.

Acemoglu, D., Antras, P. and Helpman, E. (2007), 'Contracts and Technology Adoption', in *American Economic Review*,97(3): 916-943.

Aghion, P. and Schankerman, M. (2004), 'On the Welfare Effects and Political Economy of Competition-Enhancing Policies', in *Economic Journal*,114(498): 800-824.

Alesina, R. and Zeira, J. (2006), 'Technology and Labor Regulations', *NBER Working PaperNo.12581*, The National Bureau of Economic Research, Cambridge, MA.

Arnold, J. and Javorcik, B. (2005), 'Gifted Kids or Pushy Parents? Foreign Acquisitions and Plant Productivity in Indonesia', *World Bank Policy Research Working Paper3597*, Washington D.C.

Barbosa, N. and Faria, A. (2008), 'Technology Adoption: Does Labour Skill Matter?', in *Empirica*, 35(2): 179-194.

Bartel, A. and Sicherman, N. (1998), 'Technological Change and the Skill Acquisition of Young Workers', in *Journal of Labor Economics*, 16(4): 718-55.

Baumol, W. (1990), 'Entrepreneurship: Productive, Unproductive and Destructive' in *Journal of Political Economy*,98(5): 893-921.

Brown, J., Earle, J. and Telegdy, A. (2006), 'The Productivity Effects of Privatization: Longitudinal Estimates from Hungary,

Romania, Russia, and Ukraine', in *Journal of Political Economy*,114(1): 61-99.

Chun, H. (2003), 'Information Technology and the Demand for Educated Workers: Disentangling the Impacts of Adoption versus Use', in *The Review of Economics and Statistics*, 85(1): 1-8.

Coe, T., Helpman, E. and Hoffmaister, W. (1997), 'North-South R&D Spillovers', in *The Economic Journal*, 107(440): 134-49.

Cohen, W. and Levinthal, D. (1989), 'Innovation and Learning: The Two Faces of R&D', in *Economic Journal*,99(397): 569-596.

Cohen, W. and Klepper, S. (1996). 'Firm Size and the Nature of Innovation within Industries: the Case of Process and Product R&D', in *Review of Economics and Statistics*,78(2): 232-243.

Comin, D. and Hobijn, S. (2004), 'Cross-Country Technology Adoption: Making the Theories Face the Facts', in *Journal of Monetary Economics*,51(1): 39-83.

Dasguta, S., Agarwal, D., Ioannnidis, A. and Gopalakrishnam, S. (1999), 'Determinants of information technology adoption: An extension of existing models to firms in developing country', in *Journal of Global Information Management*, 7(3): 30-40.

Doms, M., Dunne, T. and Troske, K. (1997), 'Workers, Wages and Technology', in *The Quarterly Journal of Economics*, 112(1): 253-290.

Djankov, S. and Hoekman, B. (2000), 'Foreign Investment and Productivity Growth in Czech Enterprises', in *World Bank Economic Review*,14(1): 49-64.

Djankov, S., McLieshand, C. and Ramalho, R. (2006), 'Regulation and Growth,' in *Economics Letters*, 92(3): 395-401.

Easterly, W. and Levine, R. (2001), 'What Have we Learned from a Decade of Empirical Research on Growth? It's Not Factor Accumulation: Stylized Facts and Growth Models', in *World Bank Economic Review*,15(2): 177-219.

Geroski, P. (2000), 'Models of technology diffusion', in *Research Policy*, 29(4-5): 605-625.

Green, J, Kimuyu, P., Manos, R. V. and Murinde, V. (2002), 'How do small firms in developing countries raise capital? Evidence from a large-scale survey of Kenyan MSEs', *Finance and Development Working Paper No.* 54, University of Manchester, UK.

Esfahani, H. and Ramirez, M. (2003), 'Institutions, Infrastructure, and Economic Growth', in *Journal of Development Economics*,70(2): 443-477.

Fabiani, S., Druant, M. , Hernando, I., Kwapil, C., Landau, B., Loupias, B., Martins, F., Mathä, F., Sabbatini, R., Stahl, H. and Stokman, A. (2005), 'The pricing behavior of firms in the euro area: new survey evidence', *ECB Working PaperNo. 535*, European Central Bank.

Hall, B. and Khan, B. (2003), 'Adoption of New Technology', *NBER Working PaperNo. 9730.*

Helpman, E. and Krugman, P. (1986), *Foreign Trade and Market Structure,* Cambridge: MIT Press.

Hollenstein, H. (2004), 'Determinants of the adoption of Information and Communication Technologies (ICT): An empirical analysis based on firm-level data for the Swiss business sector', in *Structural Change and Economic Dynamics,* 15(3): 315-342.

Keller, W. (2004), 'International Technology Diffusion,' in *Journal of Economic Literature*,42(3): 752-782.

Karshenas M. and Stoneman, P. (1993), 'Rank, Stock, Order and Epidemic Effects in the Diffusion of New Process Technologies: An Empirical Model,' in *Rand Journal of Economics,* 24(4): 503–528.

Lee, J. (2004), 'Discriminant Analysis of Technology Adoption Behaviour: A Case of Internet Technologies in Small Business', in *Journal of Computer Information Systems,* 44(2): 146-157.

Lefebvre L., Lefebvre, E., Elia, E. and Boek, H. (2005), 'Exploring B-to-B E-Commerce Adoption Trajectories in Manufacturing SMEs', in *Technovation,* 25(12): 1443-1456.

Love, P.E.D., Irani, Z., Standing, C., Lin, C. and J.M. Burn, J.M. (2005), 'The Enigma of Evaluation: Benefits, Costs and Risks of IT in Australian Small-Medium-Sized Enterprises,' in *Information and Management*,42(7): 947-964.

Nelson, R. and Phelps, E. (1966), 'Investment in humans, technological diffusion, and economic growth', in *American Economic Review: Papers and Proceedings*,51 (2): 69-75.

Oliner, S. and D. Sichel (2000), 'The Resurgence of Growth in the Late 1990s: Is Information Technology the Story?' in *Journal of Economic Perspectives*,14(4): 3-22.

Pavitt, K. (1984), 'Patterns of Technical Change: Towards a Taxonomy and a Theory', in *Research Policy*, 13(6): 343–373.

Parente, S. and Prescott, E. (1994), 'Barriers to Technology Adoption and Development', in *Journal of Political Economy*,102(2): 298-321.

Pavlova, A. (2001), 'Adjustment Costs, Learning-by-Doing, and Technology Adoption under Uncertainty', *MIT Sloan Working Paper No. 4369-01*, MIT Sloan School of Management, Cambridge, MA.

Porter, M. (1990), *The Competitive Advantage of Nations*, New York: Free Press.

Prescott, E. (1998), 'Lawrence R. Klein Lecture 1997: Needed: A Theory of Total Factor

Productivity', in *International Economic Review*,39(3): 525-551.

Roller, L. and Waverman, L. (2001), 'Telecommunications Infrastructure and Economic Development: A Simultaneous Approach,' in *American Economic Review*,91(4): 909-923.

Schumpeter, J. (1942), *Capitalism, Socialism, and Democracy*, New York: Harper.

Timmer, J. and van Ark, B. (2005), "Does information and communication technology drive EU-US productivity growth differentials?" in *Oxford Economic Papers*, 57(4): 693-716.

Wagner, J. (2007), 'Exports and Productivity: A Survey of the Evidence from Firm Level Data', in *The World Economy*,30(1): 60-82.

World Bank (2008), *Global Economic Prospects 2008: Technology Diffusion in the Developing World*,Washington, DC: World Bank.

Chapter 5

Testing Technology Transfer Policies for the South African Seed industry

Sebastian Derwisch & Birgit Kopainsky

1. Introduction

The yields of cereals in Sub Saharan Africa have been stagnating in the past decades. At the same time the yield gap between Sub Saharan Africa and the rest of the world has widened (The World Bank, 2008). Considering that the majority of African agriculture is subsistence and that Sub Saharan Africa is experiencing the largest urban growth in the world, increasing productivity of cultivated areas and turning subsistence agriculture into small-scale commercial agriculture is crucial for food security. The development of improved seed varieties is a knowledge intensive process of innovation generation. Many developing countries remain low research & development performers (Djeflat, 2009). One reason for this development in the agricultural input industries is the lack of investment into the commercial seed sector. With the exception of a few countries such as South Africa, there is very little evidence of successful commercial seed sector development (Tripp, Rohrbach 2000). Even if there is commercial seed sector development, domestic companies are only marginally involved. The market share of local companies e.g. in Malawi is below 5% and even in South Africa it is below 15%. From a food security and poverty alleviation perspective, however, domestic innovation would be very important. Multinational enterprises in the seed business produce new seed varieties according to their classification of mega environments and therefore only breed for a limited variety of climatic conditions. Farmers in marginalized areas are thus not able to make use of the innovations of large companies. A domestic seed industry is therefore important to provide improved seed varieties also for small-scale farmers. Moreover, the development of improved varieties that are adapted to local conditions contributes to maintaining and enlarging agro-biodiversity. An innovating domestic seed industry is also beneficial for a country in economic terms. Multinational enterprises lower domestic savings and investment by siphoning off capital through their access to capital markets and foreign exchange (Moran, 1998). This paper addresses the question why domestic seed companies in South Africa have failed to develop or maintain market share, even though commercial seed sector development can be observed.

We hypothesized that the high degree of market power of multinational seed enterprises in emerging markets is a consequence of path dependent processes that result from positive feedback loops of accumulation of resources and market power (Howard, 2009). This inhibits the emergence of an innovating domestic economy. The research questions that follow from this hypothesis are: (i) which processes lead to path dependency in the agricultural sector, (ii) how can path dependency in the agricultural seed sector be unlocked so that the development of a domestic seed industry is stimulated and (iii) which role do Intellectual Property Rights play in these processes.

For this purpose we build a dynamic simulation model of the

process of research and development (R&D) in the seed industry and calibrate it to the historical development and current situation of the maize seed industry in South Africa. The simulation model describes the processes through which the domestic and multinational seed industries develop improved maize seed varieties and the transactional structures through which the two industries have so far interacted, which has been mainly via acquisitions.

In our analysis we identified the reinforcing feedback loops that lead to path dependency. We identified how the actors in the seed market interact and the role of different resources in the process of variety development. This is necessary to develop policies that can unlock path dependency. Then we tested the effect of different policies that focus on increasing the resource stocks of the domestic sector. We introduced new possible channels of interactions between domestic and foreign companies and focus on the role of spillovers to break path dependency patterns for the domestic and the multinational seed industries in different stages of development. Spillovers from multinational enterprises to the domestic seed industry might help breaking path dependency. However, these spillovers are currently often not found as a consequence of the tight control of multinational enterprises over technology, higher management functions and export channels (Moran 1998). With this analysis we address the demand for an innovation systems approach that integrates management, development and innovation for better policy outcomes and learning (Pant and Hambley, 2009).

2. The historical development of the South African seed industry

To understand the processes of market development in South Africa, the actors and their role in the seed market we first give a historical overview of the development of the seed industry in South Africa. As an indicator of how the capacity of the multinational and the local seed industry have developed we compare the variety release rates over the past 25 years. The number of varieties released is considered a standard indicator to capture research system performance (Spielmann and Birner, 2008).

Figure 1 shows the number of maize variety released in South Africa in the past 28 years for the domestic seed industry and the multinational seed industry. Domestic variety release refers to maize seed varieties that originated from a South African company, and multinational variety release refers to maize seed varieties that originated from a company that was not or not anymore South African. Maize varieties that were released by the public sector are excluded from the analysis. Figure 1 illustrates that generally, variety release increases over time. Domestic variety creation increases until the 1990s and then shows decreasing cycles. At the same time multinational variety creation increases after stagnating at low levels before the 1990s. Due to the apartheid regime and barely present foreign market influences, the domestic seed industry had initially developed well. Pannar, the leading domestic private company held 60% of the market share in maize seed. The council for scientific and industrial research (CSIR) was very active as well, releasing varieties via Sensako, a South African seed company that was semi-public and a recipient of public materials that other domestic seed companies could license for free. When the apartheid regime ended, South Africa opened up for foreign direct investment and attracted multinational seed companies. The reinforcing feedback loops described in Figure 2 initially worked in favour of the domestic seed industry. The domestic seed industry followed a path dependent development process in the 1980s as its investments were higher than those of the multinational seed industry. The domestic seed industry was thus gaining market share. But as more foreign firms entered, they reversed that trend. The transactional structure for multinational companies to gain access to resources was acquisitions of domestic companies. In the early 1990s, Pioneer, Cargill, Syngenta and Monsanto entered the market and acquired a series of domestic seed companies. Monsanto for example, took over Sensako and thus immediately controlled 20% of the market. As a result capital, labour and genetic resources were drained from the domestic seed industry, increasing the stock of these resources in the multinational seed industry. The quantity of the resources withdrawn from the domestic seed industry in the course of these acquisitions turned the reinforcing loop around from augmenting the innovative capacity of the domestic seed industry to draining it further. The same reinforcing feed-back loop worked in the opposite direction for the multinational seed industry and led to continuous increases in its innovative capacity.

The drop in variety release by the withdrawal through acquisition for the domestic sector was larger than the increase in variety release by the multinational sector as we see in Figure 1. This is due to the resource levels that differ between these two sectors. In the initial scenario in 1978, the two sectors started with the same resource level. But as capacity creation takes less time for the multinational sector, by the time the acquisition took place in the early 90's, the resource level of the multinational sector was higher and thus the relative share of value withdrawn from the domestic sector higher than the relative share gained by this acquisition by the multinational sector.

The acquisition of Sensako by Monsanto in the year 2000 had far reaching intellectual property rights consequences. Sensako used to be a semi-private company that received genetic resources and varieties from the public sector and made it available to other domestic seed companies. That route of access to genetic resources for the domestic seed industry was closed once Monsanto bought Sensako. IPR in the form of Plant Variety Protection and contract law in the form of bag tags now prevent domestic seed companies from using genetic resources that were appropriated by Monsanto for further breeding. This considerably reduces the possibility of domestic companies to produce high quality seed as access to genetic resources of high quality is of great importance for variety development.

Figure 1: Yearly variety release rate of the domestic and multinational seed sector (1978 – 2006)

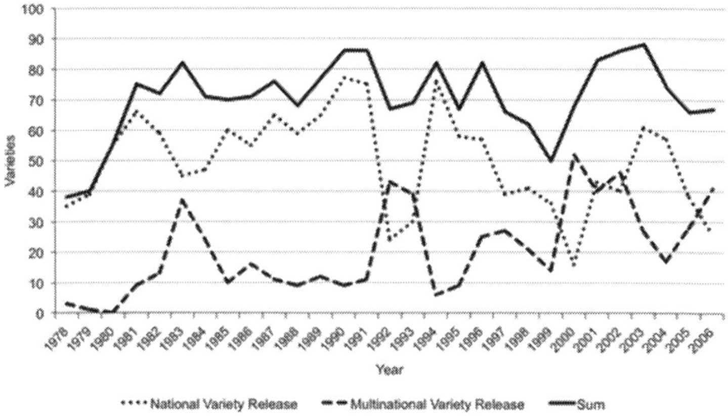

Source: Variety Release Office South Africa

3. Methodology

We use the information about the South African seed market development to establish a hypothesis of the processes that have led to the trend that we observed in variety release. We complement the literature about the South African seed market with data from interviews that were carried out with public and private actors in the in South African seed sector in 2010. A large body of literature examined the impact of foreign direct investment (FDI) on the productivity of an industry, the influence that intellectual property rights (IPR) have on the quality and quantity of technology that is transferred from multinational companies to their local affiliates and how FDI affects the development of the local economy via spillovers (Derwisch and Kopainsky, 2009). The literature mainly uses econometric analyses to study individual links between foreign ownership of companies and productivity levels (Blomström and Persson, 1983; Wamae, 2009). Models that integrate the existing body of literature and capture the dynamics in the system are largely missing (Novoselova et al., 2007). We used the existing literature to establish the links between several variables and formalize them in a dynamic simulation model.

The approach

For developing the simulation model, we adopted a system dynamics approach. System dynamics is a computer aided approach to policy analysis and design. It applies to dynamic problems (problems that involve change over time) arising in complex social, economic, or ecological systems, i.e., in systems characterized by feedback loops, delays, and nonlinearities (Richardson, 1991; Sterman, 2000). An example for a feedback loop in the context of this paper is the circular causality between variety creation and revenues, where increases in the stock of varieties result in sales and thus revenues. This, in turn enables companies to reinvest in resources for vareity development which increases variety production even further. Delays are omnipresent, also in this example of a feedback loop, as time elapses between increases in resources for variety creation expenditures and the observable increases in the variety stock. Examples for nonlinearities are the elasticities with which the production function of the variety creation rate changes in accordance with changes in the resource levels. The system dynamics approach involves the development of a simulation model. Experimentation with computer simulation models allows the design and analysis of different scenarios and policies in a risk free environment (Sterman, 1994). Dynamic simulation models are sets of equations that describe the behavior of dynamic systems. The models study cause and effect. Given specified initial conditions and assumed behavioral parameters, the models trace the changes in key variables over time and allow seeing the dynamic implications of the assumptions (Ruth and Hannon, 1997). Mathematically, the basic structure of a formal dynamic simulation model consists of a system of coupled, nonlinear, first-order differential (or integral) equations. Simulation of such systems is accomplished by partitioning simulated time into discrete time intervals of length dtand stepping the system through time one dt at a time. By breaking the simulated time into discrete intervals dtsimulation makes possible the creation and use of models that cannot be solved in closed form. Simulation thus expands the range and complexity of problems that can be modeled (Richardson, 1991).

The conceptual framework

The aim of the simulation model we build is to capture the long-term development (1980-2007) of the innovative capacity of the domestic and multinational maize seed industries in South Africa. The analysis of the simulation model helps to characterise the range of behavioural patterns that these links generate (Kopainsky and Luna-Reyes, 2008).

Seed varieties are the end product of the innovation process for private companies. The innovative capacity of seed companies can therefore be measured by the number of improved maize seed varieties released to the market. We aggregated seed sector actors into a domestic seed producing industry and multinational seed companies. With such high level of aggregation, we can focus on the analysis of how to develop a functioning innovation system, where actors of the domestic and the multination seed industries interact and thus raise the overall level of sophistication. Figure 1 shows how domestic and multinational seed industries compete for market share. Sales generate revenue, which are reinvested in resources needed to breed new seed varieties (capital, labour, genetic resources). Assuming that new varieties represent improvements, an increase in the stock of new varieties of e.g. the domestic seed industry will increase the incentive of farmers to purchase domestic seed varieties. Due to higher sales, there will be higher revenue for the domestic seed industry and more domestic investment in labour, capital and genetic resources. This will allow for an increase in the release of new domestic seed varieties after a time delay. On the contrary, a decrease of the stock of new varieties marketed of the domestic seed industry will provide for less innovation from domestic companies and thus make it less attractive for farmers to purchase seed varieties from them. This will result in fewer sales and revenues and subsequently also in less investment, resulting in a lower rate of new variety release.

Figure 2 illustrates how these relationships are integrated into a series of feedback processes. Reinforcing processes that cause increasing returns to scale are seen as an important condition for path dependency (Arthur, 1994; David, 1974; North, 1990).

Figure 2: Structure of the conceptual framework of the South African seed market

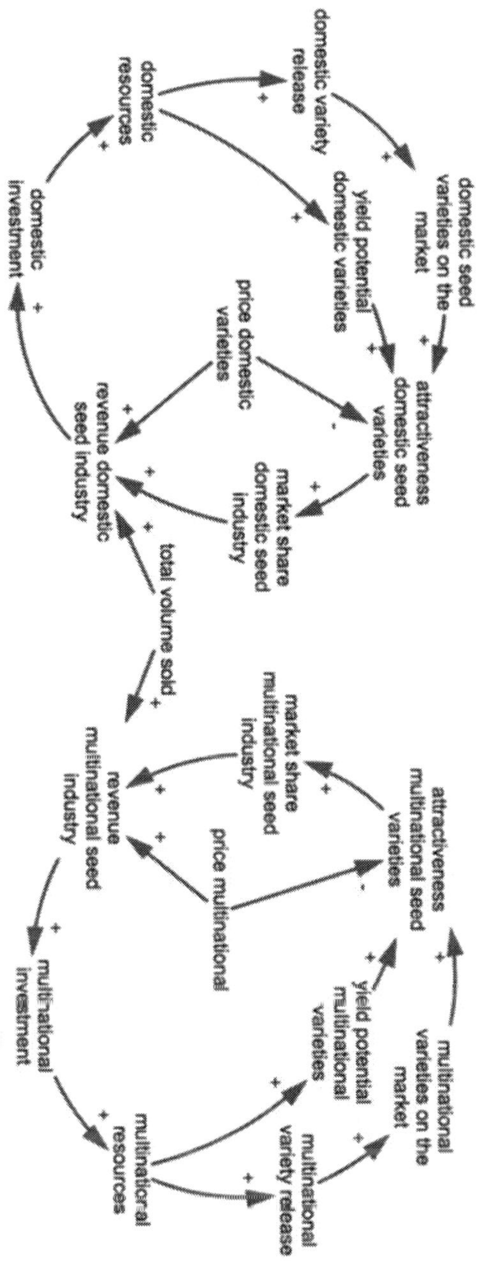

Formalization of the conceptual framework

Having established the theory for the links between the single actors and the process that result in varieties creation and release, we formalized these linkages by using differential equations. The description of the model structure in the following paragraphs will focus on the domestic seed industry. The multinational seed industry has, if not indicated otherwise, the same structural components.

We begin with the description of the innovative capacity of the domestic seed industry, which is the result of the plant breeding process. In the simulation model, it is represented as a stock. The stock describes the number of improved maize seed varieties that the domestic seed industry offers on the market each year. The total stock of domestic seed varieties on the market Vd accumulates from domestic variety creation cd less the discard of domestic seed varieties dd . Domestic seed varieties are discarded after an average of 4 years on the market, when farmers do not demand these varieties anymore. Table 1 lists the values for all the parameters in the simulation model. Equation 1 illustrates the process of variety accumulation.

$$V_d(t)= V_d(t_0)+\int[\, c_d - d_d]\, dt \qquad 1$$

Variety creation or plant breeding therefore determines how the number of improved varieties develops over time. The quality of the varieties is also a direct result of plant breeding. Quality encompasses many criteria but yield is the most important criterion in commercial agriculture. We assume that every new variety that is marketed embodies also an innovation that is usually reflected in an increase in potential yield. Thus, with the number of varieties, the yield potential also increases over time. Increase of potential yield is an exogenous variable in our model. Domestic and multinational companies have different increases in yield potential as they use different technologies. This is also represented in the fact that variety development is more expensive for the multinational sector than it is for the domestic sector. The annual yield increase rate for the domestic sector is about 0.5% whereas it is 4% for the multinational sector, (WDR, 2008). We represent the quality of varieties in terms of yield potential as a stock that is related to the variety stock. The total stock of yield potential Y_d accumulates from the rate of yield potential increase y^c_d less the rate of yield potential decrease for domestic varieties y^d_d as highlighted in equation 2.

$$Y_d(t)= Y_d(t_0)+\int[y^c_d - y^d_d] dt \qquad ......2$$

For the purpose of this analysis we aggregated the input factors that are required for breeding into three resource stocks: capital, labour, genetic resources. Into capital we summarize material input factors like the physical capital (e.g. seed processing and storage facilities) and the land that is needed. The total stock of domestic capital C_d accumulates from domestic investment rate i_d in capital less the discard rate of domestic capital d^c_d as defined in equation 3.

$$C_d(t) = C_d(t_0)+\int[i_d - d^c_d] dt \qquad ......3$$

Labour is another input factor for plant breeding. The labour stock summarizes the plant breeders with their capacity and knowledge related to the plant breeding process. (Acquaah, 2007) This process is shown in equation 4. The total stock of domestic labour L_d accumulates from domestic investment in labour i^l_d less the attrition rate of domestic labour d^l_d. The attrition rate of domestic labour is determined by the average number of years employees remain with a domestic seed company.

$$L_d(t) = L_d(t_0)+\int[i^l_d - d^l_d] \qquad ......4$$

Genetic resources are not only used for reproduction but through genetic manipulation (plant breeding) they can be improved for better performance of the crop (Acquaah, 2007). The total stock of domestic genetic resources G_d accumulates from domestic genetic resources creation rate c^g_d as shown in equation 5.

$$G_d(t) = G_d(t_0)+\int[c^g_d] dt \qquad ......5$$

The three resource stocks capital, labour and genetic resources are measured in monetary terms. The creation rate of new varieties c_d is calculated as the quotient of the initial resource level and the amount of capital needed to create one variety s_{vd}. This is multiplied by production function that is formulated according to Cobb and Douglas (1928) as the product of the three resources to the power of elasticities ecd for capital, eld for labour and egd for genetic resources, and divided by the time it takes to develop a new variety t_{vd}. The costs for developing one new variety differ for the multinational and the domestic seed sector. Creating a new variety costs about 50000 Rand per year for domestic companies and about 1.5 M Rand for multinational companies. The production function for new varieties is therefore defined in equation 6.

$$c_d = (ir_d / s_{vd})^* ((C_d)^{ecd} * (L_d)^{eld} * (G_d)^{egd})) / t_{vd} \text{ ....} 6$$

The inflows into the resource stocks (investment rate in capital, investment in labour, genetic resources creation rate) are determined by the total investment of the domestic seed industry in research and development (R&D). For seed companies, investment in R&D is around 15% of the total revenue. This overall investment is allocated to labour, capital and genetic resources according to their relative importance for variety development. As illustrated in equation 7, total revenue r_d is calculated by multiplying the price of seed p_d and the volume that is sold. The volume sold is the product of the area that is planted with improved maize seed varieties in ha a, the seed needed per ha s^n and the market share m^s_d of the domestic seed industry.

$$r_d = pd^* (a^*sn * msd,) \text{ ....} 7$$

The market share of the domestic seed industry is calculated as a ratio of the attractiveness a_d of domestic varieties over the total attractiveness a_t, which is the sum of the attractiveness of varieties of the domestic and the multinational seed industry. The attractiveness of seed varieties a_d is a concept that we use to calculate the value that the customer attaches to the seed, which is illustrated in equation 8. It is the product of the innovativeness i^n_d which represents the value that the customer attaches to the fact that the company releases new products regularly and is calculated from the stock of the number of domestic varieties on the market and the yield (i.e. potential yield stock Y_d). In our model, the price is exogenous and not subject to change as the seed market is quality and not price driven (source-interviews). Numbers of varieties and yield are assigned weights for innovativeness (wi) and yield (wy) according to their relative importance to the customer. As the seed market is less price-driven and primarily determined by the actual quality that companies offer, the highest weight is assigned to yield.

$$a_d = (i^n_d)^{wi} * (Y_d)^{wy} \quad \ldots\ldots 8$$

Data collection

Table 1: Data sources and parameter values for seed variety creation variables

Variable	Value	Data source
initial labour domestic, $L_d(t_0)$	1.85 M USD	The initial values of the resource stocks were calculated based on the costs of variety development for domestic and multinational companies. Developing one variety costs about 45000 USD for domestic companies and 1.5 Mio for multinational companies (source: expert estimates collected during field work). Approximately 70% of the total expenses for variety development are caused by labour, 20% by capital and 10% by genetic resources. These shares can, however, vary depending on the base of genetic resources that a company already has and the quantity of genetic resources it produces. Thus, by knowing the initial number of varieties that were produced in the domestic and multinational industry in the year 1978, the initial stocks measured in Rand can be calculated.
initial labour multinational	1.85 M USD	
initial capital domestic, $C_d(t_0)$	3.75 M USD	
initial capital multinational	3.75 M USD	
initial genetic resources domestic, $G_d(t_0)$	250000 USD	
initial genetic resources multinational	250000 USD	
initial domestic seed varieties on the market $V_d(t_0)$	36 varieties	Department of Agriculture, Republic of South Africa
initial multinational seed varieties	2 varieties	Department of Agriculture, Republic of South Africa

elasticity genetic resources	0.5	Expert interviews with domestic and multinational seed companies in South Africa and with multinational seed companies abroad
elasticity labour	0.3	
elasticity capital	0.2	
area planted with maize	2'800'800 ha	FAOSTAT, 2010
seed needed per ha	25 kg/ha	sourceMorris, 1998
relative weight yield wy	0.7	Expert interviews with domestic and multinational seed companies in South Africa and with multinational seed companies abroad
relative weight innovativeness wi	0.35	
Money necessary for domestic variety creation s_{vd}	50000 Rand/year	Expert interviews with domestic seed companies in South Africa
Money necessary for multinational variety creation	1500000 Rand/year	Expert interviews with multinational seed companies in South Africa
average time of domestic variety on the market	4 years	Expert interviews with multinational seed companies in South Africa
average time of multinational variety on the market	4 years	Expert interviews with multinational seed companies in South Africa
time to create domestic variety t_{vd}	6 years	Expert interviews with multinational seed companies in South Africa
time to create multinational variety	3 years	Expert interviews with multinational seed companies in South Africa
price of domestic varieties p_d	20 Rand/kg	Expert interviews with multinational seed companies in South Africa
price of multinational varieties	25 Rand/kg	Expert interviews with multinational seed companies in South Africa

Source: Own compilation of data

An important step in the model-building process is to quantify the model's parameters. One set of parameters are the initial values for the stocks. Another set are the constants in the model. Data is also used for the calibration of the simulation model. In this case, the time series for a particular variable, which is produced by the model is compared to a time series perceived in the system. Parameters with no available data source can be estimated or calibrated such that the simulated time series for other variables matches the observed data for that variable as closely as possible. Table 1 provides an overview of the data used for the specification and calibration of simulation model. Data used in the model is based on a series of interviews with managers of private maize seed companies and scientists from the public research institute as well as available statistical data for variety release obtained from the ministry of agriculture and on relevant literature (values and references are indicated in Table 1).

Policy Design

From an industrial development and poverty alleviation perspective, the current ownership patterns and market structures in South Africa are undesirable. To address the questions of how path dependency in the agricultural seed sector can be broken and how the development of a domestic seed industry be stimulated, we designed and analyzed policies that might be able to break this path dependency. The aim of the policies is to test how the resource stocks of the domestic sector can be increased to support innovation and domestic industry growth. We tested how different transactional structures between the domestic and multinational seed industry cause changes in market share and varieties on the market. This is an innovative aspect of this study since "most studies on spillover tend to simply test the existence of spillover effects but pay little attention to the channels through which spillovers happen" (Fan, 2002). Having modelled the base run of the past years, we used the simulation model to make projections into the mid term future (2011 – 2030). The scenarios we assumed related to three technology transfer possibilities, which are illustrated in Figure 3.

The first policy option looks at the effect of demonstration effects. Demonstration effects are an important source of spillover (Maskus, 2000), through which domestic seed companies adopt technologies. In the case of plant breeding, demonstration effects occur through imitation or reverse engineering of genetic resources. This creates a flow of genetic resources to the domestic seed industry. Figure 3 illustrates the additional structure that was added in order to model the flow of resources from the multinational to the domestic industry (dotted line). We limit demonstration effects to the stock of genetic resources. For the case of demonstration effects, we modeled the flow of information and material as the transfer of a fraction of the stock of genetic resources from the multinational seed industry to the domestic seed industry. The magnitude of spillovers by demonstration effects is difficult to determine as there is little empirical evidence. We estimated the amount of spillover that occurs to the domestic industry via demonstration effects to be 1% of the multinational stock of genetic resources per year. The stock of genetic resources of multinational is not reduced by this form of spillover.

The second policy option is restricting foreign ownership of domestic companies. This changes the mode of interaction for foreign companies from acquisition to vertical linkages, if they want

Figure 3: **Structure of the conceptual framework of the South African seed market with new policy channels**

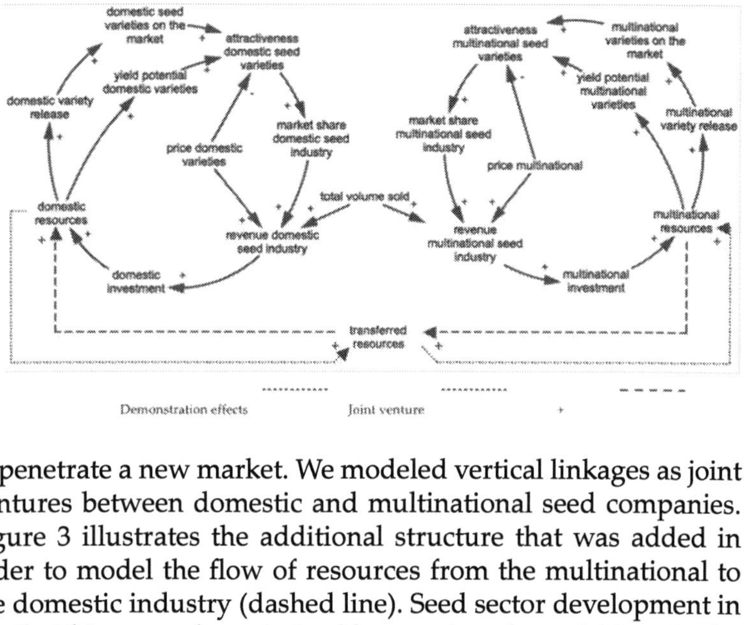

to penetrate a new market. We modeled vertical linkages as joint ventures between domestic and multinational seed companies. Figure 3 illustrates the additional structure that was added in order to model the flow of resources from the multinational to the domestic industry (dashed line). Seed sector development in South Africa was characterized by a series of acquisitions in the 1990s. To enter a new market this is the default decision for seed companies as multinationals would rather limit the diffusion of their technology in the local economy (Saggi, 2000) and as this enables companies to benefit most from the complementarities of the IP portfolios (Graff et al., 2001). However, legal frameworks can suppress the possibility for acquisitions and encourage joint ventures. In order to model the impact of joint ventures as a transactional structure, we included additional model structure. Joint ventures in the seed industry not only facilitate access to genetic resources but also to human resources (labour) and to costly production facilities (capital). We modelled joint ventures as an additional inflow to the labour and genetic resource stocks for both, the domestic and the multinational seed industry. The amount of resources exchanged depends on the length and the extent of the partnership. We set the fraction of material exchanged by a joint venture to 1% of the stock of genetic resources and labour and assumed an average duration of 10 years for a joint venture (Hewitt, 2005).

The third policy option to stimulate the domestic seed sector development is to increase investment in public sector research that then is available to the seed industry. The results of such public research would then be available for domestic as well as to the multinational seed companies. Genetic resources of high quality are expensive to develop or to license, and domestic companies do rarely have access to international networks to acquire new genetic material. Thus the development of new traits and varieties by the public sector can function as an injection of new materials that can help domestic companies to establish themselves in niches not served by the multinational companies. In our simulation we assumed the inflow of genetic resources is doubled by the influx of genetics from the public sector.

4. Results

Base case

The simulation model replicates the trends observed in historical data for the key indicators over the period from 1980 to 2006. In order to compare the model's simulation with historical data, we chose two variables for which empirical data were available: the variety creation rate of the domestic seed industry and the variety creation rate of the multinational seed industry. Figure 4 show time graphs for the two indicators comparing the model's simulation (lines labeled "base case") with historical data (lines labeled "data"). The comparison indicates that the model is not able to reproduce the oscillations observed in the data, but that it replicates the underlying trend.

Figure 4: Comparison of the variety release of the domestic (above) and the multinational (below) sector (1980 – 2010)

Table 2 contains summary statistics that describe the fit between the data and the simulation output. The root mean square percentage error (RMSE) tests the goodness of fit between the data and the simulation output. The inequality statistics decompose

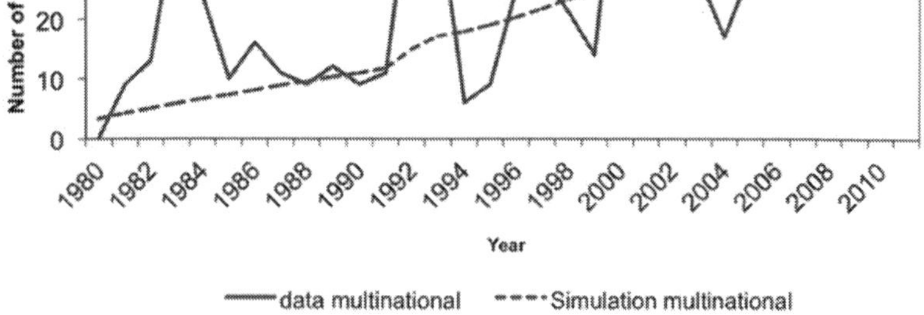

the error by dividing it into three components: the fraction of the RMSE due to bias (U^M), to unequal variation (U^S), and to unequal covariation (U^C) (Sterman 1984).

Table 2: Summary statistics for the comparison between data and model output

Variable		Inequality Statistics			
	RMSE	$U^{(M)}$	$U^{(S)}$	$U^{(C)}$	R^2
domestic variety creation rate	0.41	0.01	0.08	0.89	0.36
multinational variety creation rate	1.338	0.22	0.21	0.56	0.36

Table 2 indicates that the root mean square percentage error is fairly high for both variables and that R^2 is relatively low. The error decomposition (inequality statistics), however, shows that most of the error is due to unequal covariation U^C and thus to noise in the data. As long as the majority of the error is concentrated in U^C, the model captures the mean and trends in the data well (Sterman 2000). As shown in figure 3, the data series contains cyclical modes that are not captured by the simulation model. The purpose of the simulation model is to analyse the long-term behavior of the South African seed sector that abstracts from short term movements. Failure to capture such cycles is therefore unimportant.

Figure 1 showed the variety release rate for the domestic and the multinational seed industry. Figure 5 shows the development of the number of seed varieties on the market and thus the number of seed varieties available for the farmers. Empirical data for these stock variables were not available but the figure shows the calculation of the integral presented in equation 1. In addition to the historical period (1980 to 2010), the figure also contains the development of the seed varieties on the market for the period between 2011 and 2030 and thus the base case projections of the simulation model. The base case assumes no major shifts in framework conditions in the future compared to the historical period.

Variety production is initially higher for the domestic sector. Domestic companies have more market share, sales and revenues to reinvest. But the acquisition of domestic companies of the multinational sector in the 1990s led to a changing path dependence. Out of the three variables that in our model influence the customers' decision to buy seed from domestic or multinational companies the potential yield is the most decisive. New varieties released to the market usually represent improvements in yield.

Figure 5: Number of domestic and multinational seed varieties on the market (above) and market share of domestic and multinational companies (below) (1980 – 2010)

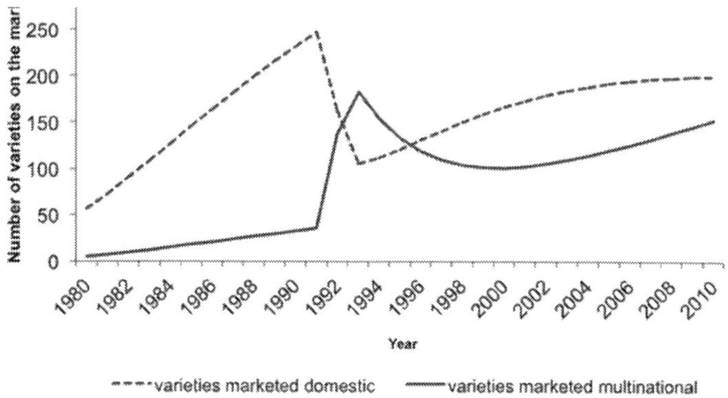

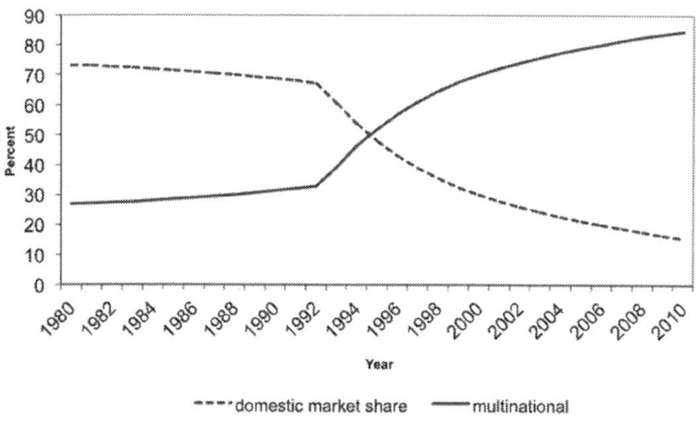

Figure 6: Impact of policies on the variety development of domestic (above) and multinational seed companies (below) from (1980 – 2030)

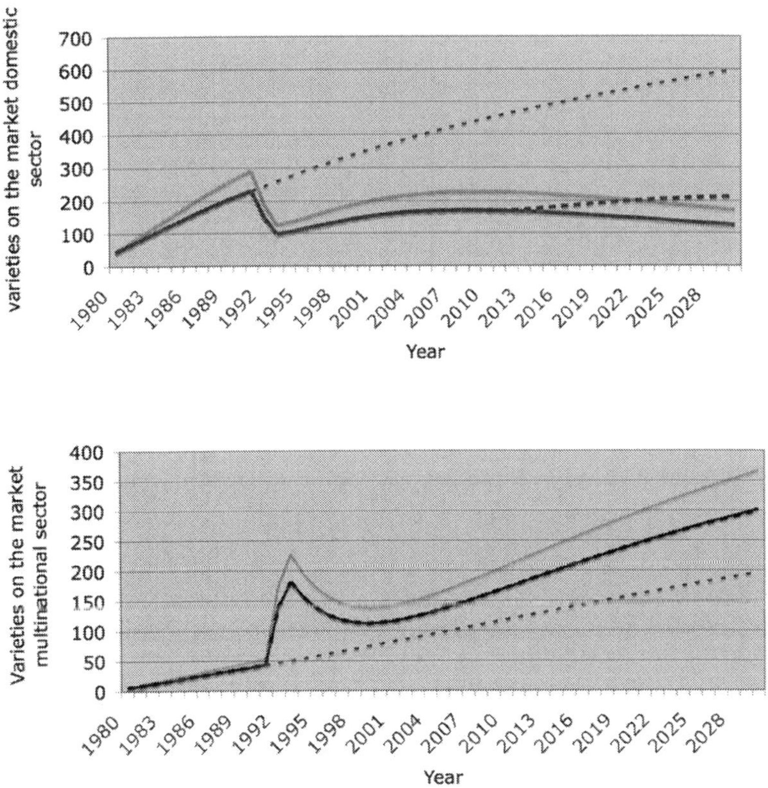

The acquisition of domestic seed companies led to a better resource equipment of the multinational seed industry, which caused a higher output with higher yield potentials that resulted in higher market shares of the multinational seed industry. Gain of market share for the multinational seed industry means loss of market share for the domestic seed industry. After 2000 then variety production is increasing at a decreasing rate whereby multinational variety production is increasing steadily. This is reflected in the market share where the multinational sector gains steadily and reaches around 85% by 2010. At this point variety production of the multinational sector has surpassed the domestic sector. As the multinational sector had acquired a large share of the genetic resources in the South African seed market and as the domestic sector was lacking an inflow of the genetic resource stock, the varieties of the multinational sector were increasingly attractive. Thus the multinational sector could gain steadily in sales and revenue whereas the domestic sector was loosing market share and sales.

Policy analysis

Variety production can be influenced by different policies. Figure 6 provides a series of policy tests with different formats of resource transfer to the domestic and the multinational companies. It compares the base run for varieties on the market to the behavior of three hypothetical situations: where there are demonstration effects, where foreign ownership of South African companies is restricted and where the public sector transfers resources.

Genetic resources represent a bottleneck for domestic companies. By demonstration effects the variety stock of the domestic sector is increased by 72 % in year 2030 compared to the base case. Due to loss of market share and revenues for multinational sector there is a minimal negative effect. Varieties on the market of multinational companies decrease by about 1 % in 2030 compared to the base scenario. In the case of the scenario in which the public agricultural research institutes transfers technology, the variety production increases for both sectors since technologies developed are public materials. Varieties on the market increase for the domestic sector by 38 % and for the multinational sector for 21 %. In the scenario of the restriction of foreign ownership share labor, capital and genetic resources are exchanged between the domestic and the multinational sector. The development of the domestic seed sector is with an increase by 388 % almost four times as high as compared to the base case. The varieties on the market of the multinational sector decrease by 35 %. Figure 5 shows that overall both sectors experience a continuous development with an overall trend to a balanced growth of both sectors.

4. Discussion and Conclusions

The global seed industry has undergone consolidation in the past decades, bringing more and more growth to fewer and fewer companies. At the same time this is constraining the growth of domestic companies in various countries through the same reinforcing feedback processes of resource accumulation. In this study we have tried to develop a deeper understanding of the dynamic trends in the interaction of a domestic industry with a foreign entrant on the level of R&D. We therefore build a dynamic simulation model that used interview data from South Africa to formalize the interaction of domestic and multinational seed companies. Our model shows that the development of the South African seed industry has been subject to path dependent resource accumulation. The existence of intellectual property rights on genetic resources has been contributing to restrict access to genetic resources for domestic companies, which has been propelling path dependent processes. Innovations, embodied in genetic resources, decide about the quality and innovativeness of a product and therefore about sales and revenues that can be reinvested in variety development.

By using the model we apply different interventions to the seed sector. Where previous studies (Aitken & Harrison 1999, Blomström et al 1983, Smarzynska & Spatareanu 2008) have focused on the static analysis of the impact of FDI on the domestic industry, we use the dynamic model to compare the effect of different policies on the development path of the South African seed industry.

The most important conclusion of our study is in line with Howard (2009) in that in order to change the development path of the seed industry, which shows a path dependent pattern, a country has to interfere in the structure of its market. Greater antitrust enforcement (Harl, 2000) or FDI restrictions that limit majority ownership of domestic companies for foreign entrants would change the pattern of interaction between the multinational and the domestic sector to joint ventures (as observed in India that has this policy applied) and enable exchanges of not only genetic resources but also skilled labor and capital. These three resources are complementary and although genetic resources represent a major bottleneck for domestic companies in

South Africa, additional skilled labor and capital are necessary to increase variety production further. Moreover is the drain of domestic capacity that happens in an acquisition scenario not possible. For this to happen the existence and the enforcement of frameworks like intellectual property rights are of crucial importance since knowledge is shared in joint ventures. For South Africa this policy option is not feasible as such regulations were not in place when the market was opened. It can however serve as an indication for a possible policy option for other countries of how to steer market development. The policy of permitting demonstration effects does increase the level of genetic resources for the domestic industry but at the same time does loose IPR enforcement discourage foreign investors and harms more sophisticated domestic companies.

The conclusion for South Africa is that the knowledge stock of the domestic industry needs to be raised or maintained. For this to happen public sector investment ist he best solution. Despite the fact that public agricultural research and development (R&D) expenditures in South Africa fluctuated considerably between 2000 and 2008, government agencies still account for the vast majority of the country's agricultural R&D spending and staffing. South Africas public sector spend on average 173 million USD yearly in the past ten years in agricultural R&D (ASTI-CGIAR, 2011). But even a considerable increase of this amount will not break the market dominance of the multinational actors but the diversity of varieties offered can be increased.

The implications of our results for policy interventions in other countries is that the South African example shows that a country needs to think early about how to steer the process of market development when it is liberalizing ist market. Interventions like public sector investment and demonstration effects in an early stage of private sector development can be used to prepare a countries private sector to the competition that a foreign entrant will represent. Another dimension of this is that for joint ventures to have a sustainable effect, the absorptive capacity of the domestic industry needs to be stimulated. A high absorptive capacity in the host country is a prerequisite for significant benefits from foreign direct investment (see Saggi, 2000). With this paper we have provided a framework to further test such conditions and to design strategies for a balanced industrial development.

References

Aitken, B. and Harrison, A. E. (1999), 'Do domestic firms benefit from direct foreign investment? Evidence from Venezuela', in *American Economic Review*,89(3): 605-618.

Acquaah, G. (2007), *Principles of Plant Genetics and Breeding*, Oxford, UK: Blackwell Publishing

ASTI-CGIAR, see: http://www.asti.cgiar.org/data/ (Accessed: March 2011)

Arthur, W. B. (1994), *Increasing Returns and Path Dependence in the Economy*, Michigan, USA: University of Michigan Press.

Blomstrom, M. and Persson, H. (1983), 'Foreign Direct Investment and spillover efficiency in an underdeveloped economy: Evidence from the Mexican manufacturing industry,' in *World Development Elsevier*, 11(6): 493-501.

Cobb, C.W. and Douglas, P.H. (1928),'A Theory of Production', in *American Economic Review*, Supplement, 18(1):139–165.

David, P. A. (1996), *Technical Choice, Innovation and Economic Growth: Essays on American and British Experience in the Nineteenth Century*, Cambridge: Cambridge University Press.

Derwisch, S. and Kopainsky, B. (2009), 'Foreign Direct Investment and Spillovers in Seed Sector Development – Towards an Assessment of the Impacts of Intellectual Property Rights,' *Paper presented at the International System Dynamics Conference*, Albuquerque, NM.

Djeflat, A. (2009), 'Building systems for innovation take off in African economies', in *African Journal Science Technology Innovation and Development*, 1(1): 54-78.

Fan, E. X. (2002),'Technological Spillover from Foreign Direct Investment – A Survey', *ERD Working Paper No. 33*, Asian Development Bank.

FAOSTAT, See: http://faostat.fao.org/default.aspx (Accessed: December 2010)

Graff, G.D., Rausser, C.G. and Small, A.A. (2003), 'Agricultural biotechnology complementary assets', in *The Review for Economics and Statistics*, 85(**2**): 349-363.

Harl, N.E. (2000),'The age of contract agriculture: consequences of concentration in input supply', in *Journal of Agribusiness*, 18(1):115–127.

Hewitt, I. (2005), *Joint Ventures*, Third Edition, UK: Sweet & Maxwell Limited.

Howard, P. H. (2009), 'Visualizing Consolidation in the Global Seed Industry 1996-2008', in *Sustainability*, 1(4): 1266-1287.

Kopainsky, B. and Derwisch, S. (2009), 'Model-based exploration of strategies for fostering adoption of improved seed in West Africa,' *Paper Presented at the International Conference of the System Dynamics Society*, Albuquerque, New Mexico.

Kopainsky, B. and Luna-Reyes, L.F. (2008), 'Closing the loop: Promoting synergies with other theory building approaches to improve system dynamics practice,' in *Systems Research and Behavioral Science*, 25(4): 471-486.

Maskus, K. (2000), *Intellectual Property Rights in the Global Economy*, Washington, D.C.: The Institute for International Economics.

Moran, D. (1998), *Foreign Direct Investment and Development: The new policy agenda for Developing countries and economies in transition*, Washington, D.C.: Peterson Institute for International Economics.

Morris, M. (ed.) (1998), *Maize Seed Industries in Developing Countries*, Boulder, Colorado: Lynne Rienner Publishers and CIMMYT.

North, D.C. (1990), *Institutions, Institutional Change and Economic Performance*, Cambridge: Cambridge University Press.

Novoselova, T.A., Meuwissen, M.P.M. and Huirne, R.B.M. (2007), 'Adoption of GM technology in Livestock Production Chains: An Integrating Framework,' in *Trends in Food Science and Technology*, 18: 175–188.

Pant, L. P. and Hambly-Odame, H. (2009), 'Innovations systems in renewable natural resource management and sustainable agriculture: a literature review', in *African Journal of Science, Technology, Innovation and Development* 1(1): 107-140.

Saggi, K. (2000), 'Trade, Foreign Direct Investment and International Technology Transfer', *Policy Research Working Paper*, T. W. B.-R. D. Group, World Bank.

Smarzynska-Javorcik, B. S. and Spatareanu, M. (2008), 'To share or not to share: Does localparticipation matter for spillovers from foreign direct investment', in *Journal of Development Economics*, 85(1-2): 176-193.

Spielmann, D.J. and Birner, R. (2008), 'How innovative is your agriculture? Using innovation indicators and benchmarks to strengthen national agricultural innovation systems', *Agriculture and Rural Development Discussion Paper no. 41*, World Bank, Washington D.C.

Sterman, J.D. (1994), 'Learning in and about complex systems', in *System Dynamics Review*, 10(2-3): 291-330.

Sterman J. (2000), *Business Dynamics: Systems thinking and modeling for a complex world*, New York: McGraw Hill.

Tripp, R. and Rohrbach, D. (2000), 'Policies for African Seed Enterprise Development', in *Food Policy*, 26:147–161.

Richardson, G.P. (1991), 'System dynamics: Simulation for policy analysis from a feedback perspective, in , Fishwick, P.A. and Luker, P.A. (eds.), *Qualitative Simulation Modeling and Analysis*, New York: Springer Verlag.

Ruth, M. and Hannon, B. (1997), *Modeling dynamic economic systems*, New York: Springer.

Wamae, W. (2009), 'Are North-South Technological Spillover substantial?A dynamic panel data model estimation,' in *African Journal of Science, Technology, Innovation and Development*, 1(1): 136–166.

World Bank (2008), *World Development Report 2008: Agriculture for Development*, Washington D.C: The World Bank.

Chapter 6

Learning, Upgrading and Innovation in Nigerian Indigenous Industries

Siyanbola Willie O. Olamade Olumuyiwa O., Isola Oluseyi O. & Egbetokun Abiodun A.

1. Introduction

In recent times developing nations can no longer compete, based only on their natural resource endowments and locational advantages. For a nation to withstand competition in this era of globalisation, there is need for such to identify its niche areas and build on it by the application of scientific methods (Egbetokun and Siyanbola, 2008). New technologies and industries may then be built around these areas of core competences. Indeed, the wealth of indigenous knowledge and technologies in Nigeria presents unique and inimitable opportunities for innovation to occur and significant avenues for growth. Oil palm in Malaysia (Abiola, 2007), herbal medicine in China and India (Acharya and Anshu, 2008; De Moraes, 2008), as well as sugarcane in Brazil (Goldemberg, 1998) are useful cases in point. Today, many cars in Brazil ride on fuels based on ethanol derived from sugarcane.

Innovation describes the use of new knowledge to offer new product or service that consumers want (Afuah, 1998). The application of that knowledge need be emphasised rather than its acquisition. In less developed countries, Lall (1991) has suggested that innovation must be thought of as a process of adoption, absorption and diffusion of available technology. When talking of innovation in a less developed economy like Nigeria, UNCTAD (2007) noted that it matters much more for countries to focus on the adaptation, diffusion and upgrading of technologies that already exist, rather than pushing (or even attempting to push) the global knowledge frontier further. It follows then, that indigenous technologies (I-Techs) offer significant opportunities for innovation in Nigeria. This is so for two main reasons: they are already available and they are already acceptable.

The relevance of technological learning to sustainable economic growth and industrialisation of any country cannot be overemphasised. It is seen as an important strategy in which 'follower' countries build up technological capability (Akerele, 2003). Akinbinu (2001) argued that technological progress, innovation and economic development, correlate with learning and not with indices such as investment in science or research and development (R&D). Sweeney (1996) noted that economies which have had consistent growth since World War II have the following characteristics: (a) the speed with which the best technical practices once known are adopted into production (this speed creates a significant gap between those using the best practice known and the practice in average use); and (b) early identification of advances in knowledge and effective utilization of the knowledge.

Akinbinu (2001) reiterated that knowing of new knowledge and information is more critical than generating new knowledge. Reich (1989; cited in Akinbinu, 2001) posited that technological leaders are those who are able to use the fruits of R&D generated elsewhere – new data, insights, inventions and prototypes – rapidly and comprehensively. Hence, small economies are not seriously disadvantaged by their inability to spend huge sums of money on R&D, provided they are receptive to new discoveries and advances in other countries and organised to make use of them (Akinbinu, 2001).

At the firm level, technological learning is described as the way in which firms acquire and build up technical knowledge and competencies and it represents the dynamic component of the process of capability acquisition (Oyeyinka-Oyelaran, 2002).

Three sub-categories of learning identified in the literature are:

i. Elemental learning: This involves "learning-by-doing" and "learning-by-adapting";

ii. Intermediate learning – Consisting of "learning-by-design" and "learning-by-improved design"; and

iii. Advanced learning which describes "learning-by-setting up a complete production system" (Lall (1989; cited in Akerele, 2003).

Oyeyinka-Oyelaran (2002) also identified seven channels of learning observable in Nigeria as follows:

i. The apprenticeship system of training;

ii. On-site training at suppliers factory;

iii. On-the job training;

iv. Expert contracting;

v. Support mechanisms provided by public institutions;

vi. Learning through transaction with local and external agents; and

vii. Learning-by-doing production and maintenance.

From the foregoing, the apprenticeship system of training is a more prevalent approach to learning among micro, small and medium enterprises in Africa. A large part of knowledge in African societies is tacit in nature. The societies thrive on oral history and most techniques are passed on from master craftsman to the apprentice. Apprenticeship lasts for between six months to three years and tends to be organised by, but not limited to, small firms (Velenchik, 1995; cited in Oyelaran-Oyeyinka, 2002).

Indigenous learning describes the acquisition of knowledge

and skills as an integral part of a specific culture. This knowledge can come from outside and be new to the culture concerned, be new but originate from within the culture, or be traditional, with a long history of being passed on within the culture. Thus, indigenous knowledge and skills are acquired through different means, knowing well that craft plays significant role in the economy of a people. Some skills are acquired through a conscious learning process, whether traditional or adapted or learned anew, whereas some crafts are learned by conscious effort rather than by doing.

Most indigenous knowledge (IK) and I-Tech are transferred in unwritten manners. To exploit them formally for innovation presents challenges, therefore. For instance, there is the problem of standardisation that has always been an issue with local technologies and practices, such as blacksmithing and traditional medicine. There is also the issue of codification which makes knowledge transfer easier and preserves its integrity over time. Meanwhile, in spite of the existence of these technologies, no appreciable effort had been made towards a comprehensive documentation of these skills and knowledge. This paper seeks to: examine some indigenous technology clusters in Nigeria, investigate mode of knowledge acquisition and transmission in the clusters; and suggest policy directions for upgrading the technologies.

2. Aluminium Casting in Saki (Southwestern Nigeria)

Background, Evolution and Technology Characteristics

Aluminium casting is a well established trade in the Southwestern part of Nigeria. It is predominant in Oyo and Osun States. The first set of practitioners introduced the trade in Saki (Oyo State) when they returned from Ghana on repatriation in 1969. Faced with unemployment and the attendant welfare problems, the returnees decided to carry on with the practice of aluminium pot casting, the technique and methodology they were exposed to and acquired in Ghana. The practice, within a short time, became so embraced in the Saki zone that the home based people became interested and enrolled as apprentices. Most of the initial apprentices had no formal education. Later, primary school leavers, either because of interest in the practice or lack of funds for further education started to show interest and enrolled as apprentices. The practice subsequently spread from Saki to other parts of the Southwest. The Saki origin of the practice was confirmed by the fact that most indigenous aluminium pots sold in Nigeria today are trade-marked with the word 'Saki' inscribed on them, regardless of where in the region they were produced.

The major raw material for indigenous aluminium casting is processed aluminium metal obtained in form of scraps from household utensils and other disused aluminium products. These scraps are melted in earthen furnaces and locally fabricated crucibles, using firewood or palm kernel shells for fuel. Molten aluminium is poured into clay moulds to form desired products. These products comprise cooking pots of various sizes. Over the years, the casting practice has evolved to the point whereby customised products for groups and individuals now emanate from the cluster. These generally include pots with capacities and shapes specified by customers. In that manner, customers have been a major source of innovation in the aluminium pottery cluster. However, due to the imperfections associated with the raw material or the production process, finished products are sometimes unsatisfactory – either leaking or too weak to withstand prolonged usage.

Bronze Casting in Benin (South-south Nigeria)

Technology Characteristics

The raw materials for bronze casting are largely metal scraps obtained both from within and outside the locality. The major equipment comprises pit furnaces and crucibles. Furnaces and clay crucibles are built by the practitioners themselves. In a very few cases, low-tonnage graphite crucibles are imported. The production technique is the lost wax method. Although the practice is not seasonal, operational disturbances may be encountered during the rainy season. The most prominent bronze casting products, some of which are shown in Figure 1, are Queen head, King head, Leopard, Standing King, Eden head, Snake head, Ife head and warriors. Depending on the sizes of the products, the time spent on production varies.

Figure 1: Benin Bronze Cast Products

Leather Tanning in Northern Nigeria

Technology Characteristics

Leather tanning is generally predominant in Northern Nigeria. The areas covered by our research are shown in Figure 2. The term tanning refers to a series of processes in which raw skins and hides are converted into more durable and benign leather products. These processes are aimed at making the skin or hide durable by making it non-putrefactive and equally yielding a product with technical characteristics suitable for a wide range of purposes. In a nut shell, tanning simply means treating skins and hides with certain chemical reagents called tanning agents in order to produce leather of high quality.

Leather is created through the tanning of hides and skins of animals, primarily goat and cattle hides. The process of tanning converts the putrescent skin into a durable, long-lasting and versatile natural material for various uses. The leather and fur industries are differentiated by the importance of their raw materials relative to the source - animals. In the leather industry, the raw materials are a by-product of the meat industry, with the meat having higher value than the skin. The fur industry uses raw materials that are higher in value than the meat and hence the meat is classified as a by-product.

Figure 2: Map of Nigeria showing the study area

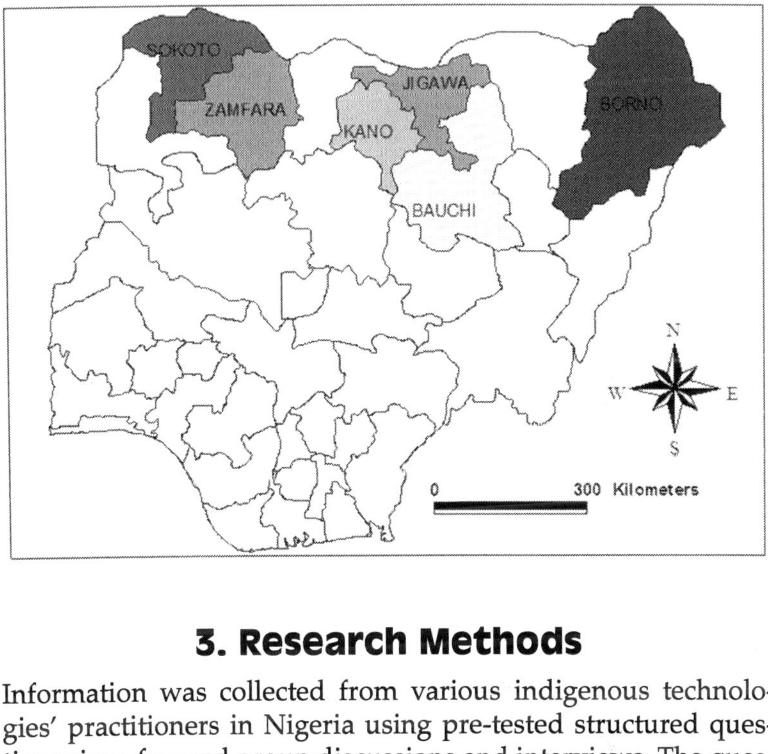

3. Research Methods

Information was collected from various indigenous technologies' practitioners in Nigeria using pre-tested structured questionnaires, focused group discussions and interviews. The questionnaires elicited information about the description of trade, equipment/machines used by the practitioners, mode of skill acquisition and innovation capability. In all, 634 questionnaires were administered (230 in the aluminium casting cluster, 104 in the bronze casting cluster and 300 in the leather tanning cluster). Of these, 599 were retrieved and analysed (213 in the aluminium casting cluster, 104 in the bronze casting cluster and 282 in the leather tanning cluster). In all, about 95% response rate was achieved. This was aided by the fact that the responses of the practitioners were recorded by the field researchers.

The discussion focuses on aluminium pottery in Southwestern Nigeria (notably Saki in Oyo State), bronze casting in the South-South (notably Benin in Edo State) and leather tanning in the Northern parts of the country (notably Sokoto, Zamfara, Kano, Jigawa, Bauchi and Borno States) (NACETEM, 2010a; NACETEM, 2010b). Specifically, we briefly describe the technological learning and capability acquisition in the clusters, as well as opportunities for upgrading and potentials for innovation.

4. Discussion of Results

Aluminium Casting in Southwestern Nigeria

(i) Learning and Capability Acquisition

Nearly all the practitioners acquired their knowledge informally, that is, without formal documentation but through oral instructions, observations and learning by doing. They claimed that if an apprentice could focus very well on the job, within a short period, such could be trusted with casting work and would be able to deliver good product. Some apprentices with good focus can acquire the skill within one year of apprenticeship, but masters often hold on to them for as long as three years. There is a hidden advantage here for the masters; the apprentices provide free labour to the masters all through their period of apprenticeship.

Observations from the field revealed that the trade is practiced in groups; thus, learning-by-doing is enhanced. Casting is not easily done by one individual performing all aspects of the work. People get together in the foundry and do several distinct works towards producing finished aluminium products. Where a single individual handles every aspect of the casting work, the process becomes very slow and number of products produced per day or week greatly reduced compared to a situation where practitioners are organized into teams.

Majority of the practitioners agreed that it will take at least an average of 3 years to learn the profession without documentation. Ninety two percent of those interviewed believed that anybody who is interested can learn the trade, 6% believed that people with some special abilities are better candidates to learn the trade while a small percentage of them (1%) opined that people of specific extraction are those who can learn the trade effectively. It is obvious here that the trade is open to parties who are interested and that participation does not require some special abilities to learn. It is to be noted that the ease of learning is conditioned by the knowledge form. If the knowledge surrounding the trade remains largely tacit, it will be unnecessarily difficult to acquire and master the trade.

It was discovered that a large proportion of the practitioners are not educated beyond the elementary level. The capability to cast advanced products (such as engine blocks, cylinder heads, hosing and other similar enclosures), which their method of production can really produce, is therefore absent. In view of the low level of practitioners' education, there was no evidence of interactions with tertiary institutions or knowledge centres that can help in improving skills as well as product and process development. This is in spite of the fact that one of the campuses of a frontline polytechnic is situated within the same locality as the pottery cluster.

(ii) Innovation/ Upgrading Challenges and Potentials

In terms of innovation, the areas of profound changes within the last 10 years are in the types of products and distribution network. Nothing much was added to method of production, no new raw materials are being used. The majority of the facilities have not been modified in the last 10 years. Those who modified their facilities have incorporated a locally made fan to blow the charcoal or the firewood they use for heating. Some of them have produced new products, but they are still products meant for cooking and these are mostly coal pots, kettles, frying pan and spoons, though the products were well received by the customers. A major product innovation observed is the production of simple electric cooker. This is considered major as a result of the low level of education of the practitioners. Also, the ability to include electrical components into products should be considered innovation for indigenous aluminium pottery makers. Many of the practitioners believed that the quality of their products is better today than 10 years ago due to better procedure.

Though existing practitioners would like to continue with the trade, there is a general concern about the future of the trade as new apprentices are now hard to find. In fact, the government poverty alleviation measure that gives easy access to motor cycles and tri-cycles as commercial means of transportation was said to have further accentuated the problem of apprentice scarcity. This is because young persons who otherwise would consider apprenticeship as an option now prefer the apparently easier money-making ventures like commercial motorcycle riding. Hence, there is a need to upgrade the trade in order to enhance its competitiveness.

On the possibility of widening the practice through training of people who are interested in the trade, 95% of the practitioners agreed to train others using informal methods that were employed to train them. Indeed, almost all the practitioners agreed to upgrade their skills if given the opportunity. They were also of the view that government mediated small training groups could be set up in different localities, if the setting up of formal schools will take a longer time to be realised. They indicated willingness to instruct in these small training groups.

About one tenth of the practitioners were found to have, at one time or the other, discontinued the practice for between 2 to 5 years at a stretch. The reasons adduced for the break include low demand, health-related problems and inadequate capital. The break is understandable because the trade is open to competition from other producers scattered all over the Southwest. Moreover, ready substitutes can be found for certain product ranges including the small household pots and spoons. Improving capabilities to the extent that products apart from pots and spoons originate from the cluster would have added more economic value to the profession and the locality. The seemingly preferential focus on the relatively simple pots and spoons is, however, not pre-conditioned by low capabilities alone. There is a general preference for aluminium pots and spoons within the Nigerian social context, especially in the Southwest where large-scale cooking for social activities like wedding ceremonies or other big social events is common. A competitive edge over industrial aluminium products exists in this regard.

Certain specific opportunities for the growth of the trade

were identified by the practitioners. These include improved methods of production, increasing demand, good export potential, adequate water supply and absence of close substitutes for most products as the top five opportunities for growth and profitability. Aluminium products from the industrial sector are not regarded as ready and available substitutes to the indigenous products. One may infer from here that the practitioners have been able to define their markets and create a niche that the more refined industrial products cannot satisfy. Limited exportation of these products currently takes place especially to the West African coast.

Table 1 shows some of the assistance that the practitioners demand from government to improve the trade. Practitioners ranked as most important assistance the mechanization of the process to make the work safer, less tedious and more productive. This is followed by improved access to raw materials and availability of micro-credit facilities. Others include access to trade fairs / exhibition, exposure to international markets, and improved power supply.

Table 1: Expected government assistance to further improve the trade

Expected government assistance	% of Practitioners
Mechanisation of the process	100.0
Improved access to raw material	99.0
Micro credit facilities	99.0
Exposure to international market	98.0
Access to trade fair exhibition	91.4
Improved power supply	84.8
Improved water supply	59.1

The role that modern science and technology can play in the aluminium casting cluster is not to be overlooked in upgrading the cluster. Such efforts as that made by the National Agency for Science and Engineering Infrastructure (NASENI) in reverse-engineering a 100kg rotary furnace is to be acknowledged and deployed within the cluster. The challenges associated with melting the metal would then be largely overcome. Although the rotary furnace is relatively expensive and not affordable for majority of the aluminium casters, the cluster can be incorporated into the recent efforts of NASENI to produce similar furnaces with lower capacity and cheaper costs for immediate deployment in micro- and small-scale industries. Government's role within this context will involve the provision of funding and equipment as well as capacity building. This could be done through organised associations, co-operative societies or micro-finance banks. Support could also come from government in the form of providing organised sources of raw materials and greater access to markets, particularly outside Nigeria.

Bronze Casting in Southsouthern Nigeria

(i) Sources and Characteristics of Practitioners

Unlike aluminium casting in Southwestern Nigeria, it was observed that less than 4% of the sampled bronze casters were born outside the traditional "Igun" bronze casting lineage of Benin. This is conditioned by the existence of a restrictive requirement of naturalisation into the bronze casting family by 'outsiders' which makes the trade less attractive to non-Igun family members, who may not want to abrogate their own family identity. The naturalisation requirement serves as entry barrier to preserve the economic appeal of the trade for the benefits of members of the traditional bronze casting family. In spite of the observed entry barrier to the trade, bronze casting holds good potentials for employment generation especially among the youth in the area.

(ii) Learning and Capability Acquisition

Learning in the bronze casting cluster occurs mainly by doing, and knowledge is transferred through two major routes. The primary route of knowledge transfer is by culture and tradition which is associated with birth or naturalisation into the Igun family. The secondary mode of knowledge transfer – though not prevalent - through which non-Iguns acquire capabilities is through apprenticeship. The strong appeal of the trade to the youth as revealed by the age distribution of practitioners suggests a competitive strength in favour of the trade. Practitioners being largely in their most productive years are likely to bring this advantage to the trade in terms of creativity, youthful strength, and productivity. Though rooted deep in culture and tradition, the predominance of youth in the trade is likely to affect the nature and pace of contact with modern technology as youths are more likely to be receptive to new ideas and experimentation. The age distribution advantage when combined with increasing entry of practitioners with above secondary level education will further boost the appeal of the trade.

(iii) Innovation/ Upgrading Challenges and Potentials

The Bronze casting trade has been sustained over many centuries, largely by being rooted strongly in the culture and tradition of the people. Nevertheless, it has wielded great economic potential to renew and sustain the interest of the present generation of practitioners. When asked about the major strengths of the trade, practitioners cited the top five strengths of the trade as its strong cultural root (88%), good export potential (86%), and increasing interest of the young to continue the trade (63%), absence of close substitute competing with Benin bronze arts (53%), and increasing local and international demand for products (48%).

The greatest challenge facing the trade is the lack of adequate capital. Most of the practitioners indicated that inadequate finance was the reason they could not purchase new equipment, improve their processes or export their products in spite of the products international appeal. Equally very important are the challenges of lack of support from government (98%), inadequate water supply (98%), erratic power supply (97%) and shortage of high quality materials (32%).

Enhancing the competitive value of this I-Tech cluster rests on two distinct needs. In the first instance, there is an urgent need to preserve the integrity of the knowledge surrounding the tech nology by ensuring that the knowledge is codified. At present, despite the high level of artistic and tourism attention that the cluster has generated, not much has been done to systematically transform the underlying knowledge from its largely oral form to codified forms. Besides the preservation of knowledge, such efforts would remove entry barriers and encourage increased participation in the trade. Knowledge centres, government and the private sector have significant roles to play along these lines. Secondly, the deployment of modern S&T is required. As with the Saki cluster, NASENI's rotary furnaces would also find applications here as much as the need for an organised source of raw materials. The need for government to show commitment through the provision of adequate infrastructure, funding and access to market goes without saying. Again, these would be greatly facilitated in the presence of organised associations, co-operatives or training schools.

Leather Tanning in Northern Nigeria

Indigenous Leather Tanning in Northern Nigeria (Figure 3) is popularly referred to as "Jima". It is done by vegetable tanning involving the use of "Bagaruwa" pods (Acacia nilotica). This is a popular plant among local tanners in Northern Nigeria. It was reported that the plant had been used in Egypt as a tanning agent between 2000 – 1570 B.C. The plant is endowed with properties which have good coloration and rapid penetration on leather. It also softens leather very quickly and with ease. The only product of the indigenous leather tanning trade is processed leather from the skin of various animals, such as goat, sheep, cow, tiger, snake, crocodile, lizard and other animals. When asked to rank in order of importance, the skins most processed were based on availability, ease of processing and marketability. The ranking in Figure 4 emerged, with sheep, goat and cow skins topping the list.

Figure 3: Tanned leather in Kano

Figure 4: Animal skins most processed

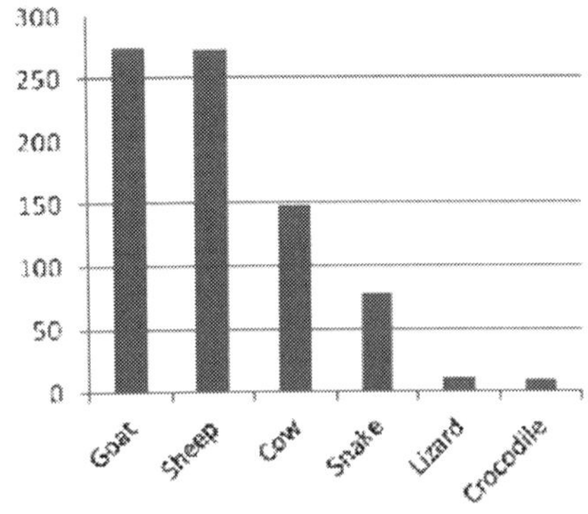

(i) Sources and Characteristics of Practitioners

About 86% of respondents claimed that leather tanning is a family tradition. Thus their involvement in the trade has its root in the knowledge of the trade being well established in the family for many generations. Others held that their involvement in the trade is based on personal interest, independent of family attachment. Respondents overwhelmingly agreed that any interested person could learn the trade. Close to 90% of practitioners claimed that they willingly chose the trade and were not forced into it as a result of family attachment or any other factor.

(ii) Learning and Capability Acquisition

The mode of knowledge transmission in leather tanning is informal as learners take oral instruction; and capability is acquired through practice. About 97% of practitioners learnt the trade this way. Because a large percentage of apprentices have leather tanning as a tradition in their family, they start learning as help hands, learning by instructions, by observation and learning to perform increasingly difficult tasks in the process. Apprentices that learn as members of practicing families generally spend less time in apprenticeship. About 76% of those who responded to the enquiry on length of apprenticeship (51% of total respondents) claimed they acquired the knowledge in less than three years. Only 3% claimed that written materials formed part of their training.

Contrary to what was observed in the indigenous aluminium and bronze casting clusters, the knowledge system around leather tanning has evolved to become much more systematic, with advanced supporting technologies. Specifically, there are two federal research institutes supporting leather technology education and practice in Nigeria. While the National Research Institute for Chemical Technology (NARICT) carries out advanced research in chemicals and leather, the Federal College of Chemical and Leather Technology (CHELTECH) provides courses of instructions, training and research in chemical, leather and leather products technology. Due to the activities of these institutions, the IK of leather tanning is now largely codified and S&T applications in the industry are pronounced.

(iii) Innovation/ Upgrading Challenges and Potentials

In spite of the apparent successes, a number of specific challenges were identified that hinder the growth of the cluster. First, little or no interactions exist between the formal institutions and the local leather tanning practitioners. Where such interactions exist, it does not include the formalisation of the apprentice training system in the local leather tanning system. Secondly, although the two formal institutions identified earlier are located within the leather tanning cluster, there is no appreciable contact between graduates of these institutions and the local practitioners. Graduates of the formal sector are largely employed in factories and rarely practice as entrepreneurs among the local leather tanning population. Thus, the influence of these institutions especially on the knowledge transmission and apprenticeship systems of the local practice is still largely limited.

5. Upgrading, New Frontiers and Opportunities for SMEs: Policy Recommendations

As presented so far, the indigenous knowledge and technologies sector in Nigeria is populated with micro- and small businesses. Indeed, given the points that have been raised In this chapter, particularly those relating to learning and capability acquisition, it is difficult to immediately see how the clusters can evolve to the stage that large enterprises would evolve from them. Nonetheless, we posit that if the issues that have been raised in the preceding sections, and subsequently herein, are addressed, medium-sized enterprises would begin to emerge from these clusters in the near future.

For instance, the world footwear and leather products sector is gradually moving from high-cost industrialised countries to low-cost developing countries. Although many developing countries, Nigeria inclusive, possess strong potentials with respect to raw material and human resources, they have mainly remained suppliers of raw and semi-finished products (Amakom, 2006). This has been attributed to a limited number of African companies and local industries having developed the capacity to produce finished products. This implies that if efforts are geared at solving this problem, Nigeria will favourably compete with some other major exporters of leather and other I-Tech products. Amakom (2006) suggested that efforts towards achieving this feat will provide partly a solution to the call for the diversification of Nigeria's economy and consequently increase the country's foreign exchange generating capacity and reducing unemployment.

The role of institutions is critical. Strong research, development and brokerage institutions are required to move the IK and I-Tech sector forward on the learning curve, assist in the codification of knowledge and facilitate product development and standardisation. The role played by the National Innovation Foundation in India is of note in this regard.

6. Conclusion and Recommendations

In this chapter, we argued that indigenous knowledge and technologies represent a significant opportunity for the attainment of national innovativeness and competitiveness. Apart from being unique and inimitable, IK and I-Techs could be readily exploited for SME development. The challenges that are presently encountered in maximising these opportunities include the informal nature of knowledge transfer and low skills levels among the practitioners. We have identified specific policy issues and directions for action to redress the influence of these challenges. Taking these together with the potentials and challenges of the Nigerian IK and I-Tech sector as discussed in the preceding sections, we make the following recommendations:

- There is the need for government, particularly at the grassroots, to acknowledge and support I-Tech development. Several ways by which this could be done includes the creation of specialised schools for capacity development; creation of specialised markets which would serve as product outlets and possibly evolve to secure international interest, the recognition of outstanding individuals in IK and I-Tech development and promotion; and the provision of venture funding.
- Institutions are identified as critical; thus the existing ones, especially research institutes, need to be strengthened while new ones are created for well-defined purposes. The activities of existing institutions like NASENI, NARICT and CHELTECH are acknowledged; but it is hard to identify other institutions focused on other identified I-Tech areas.
- Technology Business Incubators (TBIs) are important especially for business development. Lewis (2002) had actually shown that Technology Incubation works in achieving the goals of technology commercialisation by facilitating the growth of technology-based entrepreneurial firms. Thus, government at all levels as well as entrepreneurs need to demonstrate much greater commitment to locating TBIs in rural areas and city centres.
- In the area of capacity building, existing tertiary institutions and research institutes located around IK and I-Tech clusters could play the role of creating specific training programmes that could be created for the I-Tech practitioners to facilitate knowledge transmission. The institutions could also assist in upgrading and formalising the current apprenticeship system of training.
- Knowledge centres, especially tertiary educational institutions located around identified IK and I-Tech clusters have a significant role to play. Researchers from such institutions

could be involved in mapping the IK system, and help in the codification of knowledge, standardising procedures, improving processes and facilitating knowledge transfer. Students of such institutions, particularly those undertaking related studies (and undergraduate or postgraduate research) could be made to work with IK and I-Tech practitioners for stipulated periods. Besides the benefit of knowledge transfer, such arrangements would encourage practitioners to embrace formal knowledge and improve their skills levels.

In spite of the foregoing, there is still much more to be said or done about the indigenous knowledge and technologies system in Nigeria. The extensive work done by the National Centre for Technology Management (NACETEM) is only a first step in understanding a very broad landscape; further analyses relating to mapping the IK and I-Tech systems are desirable. Also, deeper understanding of skills acquisition methodologies is necessary. This would facilitate the upgrading of capacity building within the system. On a final note, we believe that the recommendations made herein, while they are far-reaching and based upon sound qualitative and empirical observations are not exhaustive. As our understanding of the IK and I-Tech system increases with time, more lessons are to be learnt and much more ideas would emerge.

References

Abiola, B. (2007), 'Divergent Paths of Development: Innovation Systems of Nigeria and Malaysia', *Guest Seminar, National Centre for Technology Management,* Obafemi Awolowo Univeristy, Ile-Ife.

Acharya, D. and Anshu, S. (2008), *Indigenous Herbal Medicines: Tribal Formulations and Traditional Herbal Practices,* Jaipur- India: Aavishkar Publishers.

Afuah, A. (1998), *Innovation Management,* New York: Oxford University Press.

Akerele, W.O. (2003), 'Learning and Technological Knowledge Acquisition in Nigeria's Small and Medium Enterprises – The Case Study of Manufacturing Enterprises in Oyo State',

NISER Monograph Series No. 7, Ibadan: Nigerian Institute of Social and Economic Research (NISER).

Akinbinu, A. (2001), 'Technological Learning in the Nigerian Passenger Car Industry: An Assessment', *NISER Monograph Series No. 12,* Ibadan: NISER.

Amakom, U. (2006), *Export constraints of the Kano leather industry: a study of industrial clusters in Nigeria,* Awka, Nigeria: Nnamdi Azikwe University.

De Moraes, M. A. F. D. (2008), 'Sugar Cane Sector in Brazil: labour indicators and migration', Available at: http://migration.ucdavis.edu/cf/files/conference_may_2008/azanha-sugar_cane_sector_in_brazil-labor_indicators_and_migration.pdf (Accessed: 30 June 2009).

Egbetokun, A. A. and Siyanbola, W. O. (2008). 'Reliance on Natural Resources is Africa's Undoing', in *The African Executive,* Available at: http://www.africanexecutive.com/modules/magazine/articles.php?article=3030 (Accessed: 30 June 2009).

Goldemberg, J. (1998), 'A guest essay: What is the role of science in developing countries?', in *Chemistry International,* 20(6): 168-170.

Lall, S. (1991), 'Human Resources, Technology and Industrial Development in Sub-Saharan Africa', in Chhiber A. and Stanely F. (eds.), *Economic Reform in Sub-Saharan Africa,* Washington D.C.: World Bank.

Lewis, D. A. (2002), *Does Technology Incubation Work? A Critical Review of the Evidence,* Ohio: National Business Incubation Association Publications.

National Centre for Technology Management (NACETEM) (2010a), 'Indigenous Technology Mapping and Analysis of Skills Acquisition Methodologies in the Aluminium and Bronze casting Industry in Nigeria', *Policy Research Project,* Federal Government of Nigeria.

National Centre for Technology Management (NACETEM) (2010b), 'Indigenous Technology Mapping and Analysis of Skills Acquisition Methodologies in Nigeria: Leather Tanning', *Policy Research Project,* Federal Government of Nigeria.

Oyelaran-Oyeyinka, B. (2002), 'Learning and Local Knowledge Institutions in African Industry', *UNU-INTECH Discussion Paper 02*, United Nations University, Maastricht, Netherlands.

Sweeney, G. (1996), 'Industry, Learning and Experience Curves: A review of Empirical Evidence', *Mimeo*, Brighton, UK: Science Policy Research Unit, University of Sussex.

United Nations Conference on Trade and Development (UNCTAD) (2007), *The Least Developed Countries Report 2007: Knowledge, Technological Learning and Innovation for Development*, Geneva and New York: United Nations.

Acknowledgements

The authors gratefully acknowledge the support and contributions of the other members of NACETEM's indigenous knowledge research team that originally implemented the projects: Indigenous Technology Mapping and Analysis of Skills Acquisition Methodologies in Nigeria. We thank specially Dr. O. A. Adetunji of the Materials Science and Engineering Department, as well as Dr. A. Y. Fasasi and Dr. A. D. Pelemo of the Centre for Energy Research and Development, Obafemi Awolowo University, Ile-Ife who were involved in the study of the Aluminium and Bronze Casting Industry. The studies were carried out with funding from the Federal Government through the Federal Ministry of Science and Technology. The support of Prof. Turner T. Isoun, the Honourable Minister of S&T at that time is gratefully acknowledged.

Chapter 7

Natural Gas in Future Nigerian National Industrial Development- An Overview in the Context of Current Progress in Technology

Aminu Bayawa Muhammad, Abdullahi Abdu Zuru, Muhammad Gidado Liman & Ahmad Ahmad Tijjani

1. Introduction

Three basic parameters could be identified as necessary for industrial development of any nation and obviously the absence of any one incapacitates the remaining. The parameters include energy, raw materials and human resource. Raw material is the substance converted to consumer-desirable products and energy is required to initiate, sustain or control the conversion process. Human resource, on the other hand, embodies the art and technological ability/expertise necessary to achieve the conversion and presentation of the product to the consumer in an acceptable/desired form. Few nations are blessed with all these parameters and fortunately Nigeria is one of them.

Nigeria is endowed with large amounts of fossil fuels, most importantly crude oil, coal and natural gas. Ironically these, predominantly energy sources, are also the most important primary raw materials for the chemical industry (Weissermel & Arpe, 2003). In other words, the fossil fuels are double blessings with respect to industrial development of the nation.

The use of coal (coal tar) as source of intermediates in dyestuff, pharmaceutical and allied industries during the industrial revolution is a historical fact. Currently, however, petroleum is the primary source of such valuable base chemicals and intermediates as ethylene, propylene, aromatics etc. Natural gas, on the other hand, is predominantly used as fuel and, worse so, in Nigeria it is predominantly used to pollute the environment as over 30% is flared.

The components and by-products of the utilisation of fossil fuels are pollutants and greenhouse gases that contribute immensely to global warming (Harrison, 2001b). Crude oil, oil shale, oil sand and coal are environmentally the dirtiest fuels. They contain higher proportions of carbon relative to hydrogen, in addition to nitrogen and sulphur, making them more prone to form CO_2, NOx, SOx and carcinogenic PAHs. On the other hand, though natural gas often contains gases such as CO_2, H_2S etc, its gaseous nature makes it fairly easy and relatively cheap to clean and therefore environmentally friendlier.

Although the predominant use of natural gas is in power generation and heating, it also finds applications in almost all other areas where petroleum products currently dominate. Moreover, natural gas is an attractive source of all base chemicals and intermediates that are currently predominantly sourced from petroleum and petro-products. These facts coupled with declining crude oil reserves, increasing cost of processing heavy crudes and high cost of processing coal (and other fossil fuels) make natural gas a very important and attractive fuel and primary raw material worldwide and particularly in Nigeria.

Here we discuss the ongoing advancement in natural gas utilisation and conversion technology with a view of having an insight into the future role it will play in industrial development of Nigeria not only as a fuel but also as an important raw material for the chemical industry of the nation

2. Methodology

The data used in this chapter were obtained from reliable sources including the government and non-governmental agencies responsible for collection of the data. Peer-reviewed articles and published books were also used as source of information. The various sources are properly acknowledged in appropriate sections of the text.

3. Natural Gas in the World Energy Scene

The world energy market is dominated by the fossil fuels which account for over 80% with oil carrying the lion's share of 36% and natural gas only 21%. However, coal, with market share of only 24% is the most abundant with reserves of about 910 billion tonnes expected to last for over 200 years at the current production rate (Copley, 2004; Harks, 2006). Over 40% of the world's natural gas proved reserves is in the Middle East. Africa has a share of only about 7.6% which is equivalent to 13.010 TCM (Figure 1 shows Africa being fourth in ranking). Of this amount about 39% (equivalent to about 5.1 TCM or 0.03% of world proven reserves) is located in Nigeria (Figure 2) (Aproot, 2004).

Over 70% of the Nigerian's economy relies on revenue generated from petroleum production. Unfortunately, according to World Energy Council, at the current rate of production, the Nigerian oil proved recoverable reserves are expected to last only about 30 years (Aptroot, 2004). Natural gas however may last much longer than petroleum; up to over 100 years (Aptroot, 2004). Important still is the fact that, natural gas can be used in almost all applications where oil-based fuels are now dominant.

Figure 1: Relative Continental Distribution of Natural Gas Deposits across the World (in BCM)

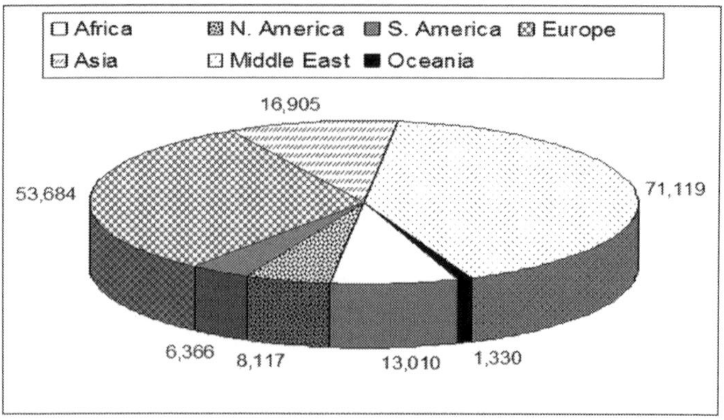

Source: Aptroot (2004)

Figure 2: The relative distribution (%) of African natural gas deposits – Selected Countries

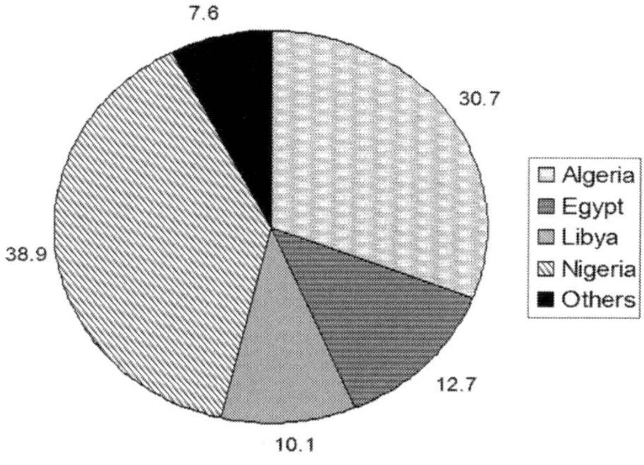

Source: Aptroot (2004).

There are many factors pushing the future energy momentum from oil towards natural gas. About 65% of the world's oil and natural gas liquids proved recoverable reserves are located in the Middle East (Figure 1). With the political instability in the region, consistent oil supply and stable price are far from being assured; Gulf war and the recent Lebanon – Israel war are living examples. Furthermore, new discoveries of crude oil are inclining towards heavy oil and this means increasing investment cost in producing and processing and thus correspondingly higher price of petroleum products. For example, the CAPEX for a new hydrotreatment unit (with a capacity for processing 12, 000 to 14, 000 barrels per day), necessary for processing heavy crudes, is between $18, 000 and $20, 000 per barrel (Moulijn *et al.*, 2001). These are in addition to the environmental problems associated with production, processing and utilisation of both coal and petroleum.

3. Natural Gas as Fuel

"Natural gas is fast becoming the economic and environmental fuel of choice" (Cook, 2006). As outlined above natural gas accounts for about 21% of world energy supply and its demand is on the increase mainly due to population and economic growth in Asia and Latin America (Isenberg, 1999; Aptroot, 2004). Most of natural gas produced is used as fuel.

More or less directly, natural gas is used for both industrial and domestic heating and electricity generation. It is also used as alternative fuel for gasoline and diesel. In domestic heating natural gas heating has been found to cost 30% less than electric heating. In the US, for example, 52% of all US homes and 70% of all new homes are heated or cooled with natural gas (NaturalGas, 2004c). Nigeria has a lot to learn from this. The encroachment of the Sahara desert has been shown to be accelerated by felling down of trees for use mainly as firewood (the predominant domestic fuel in the country). Thus, any afforestation programme in the country will surely fail unless an alternative fuel is provided. The supply of natural gas for domestic heating could be such alternative not only in increasing the efficiency in our utilisation of energy but also in arresting the desert encroachment.

Figure 3: Nigeria's Annual Natural Gas Production and Flaring (1965 to 2009)

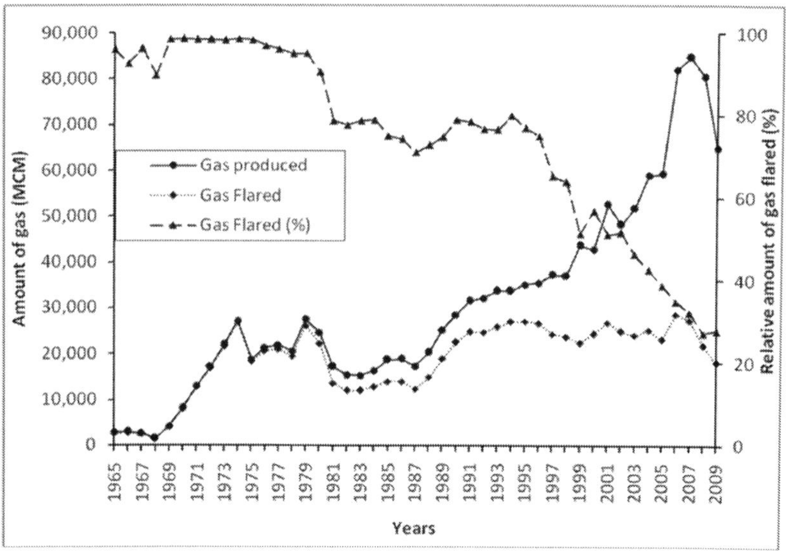

Source: CP&S (2009).

Nigeria produces an enormous amount of natural gas (up to 64,882.86 MCM in 2009). Significant amount of this is however flared in spite of general decline in the relative amounts of the gas flared over the years (Figure 3). From 1965 to 2009 about 61% of the cumulative natural gas produced was flared. Within the same period the lowest relative amount of gas flared (*ca* 27%) was recorded in 2008 & 2009. In absolute terms, however, this is equivalent to about 21,811 and 17,988 MCM, respectively (CP&S, 2009). And in monetary terms, it is worth US$52 million to 1.6 billion at the government's price bands for domestic consumers (Ashby & Ahemba, 2008).

In electricity generation, coal is the cheapest fuel. It is nevertheless environmentally the dirtiest hence the increasing use of natural gas. In 2000, for example, 95% of the 23,453 MW of electricity newly added in the US national grid was based on natural gas (NaturalGas, 2004a). This trend in use of natural gas is due both to its environmental friendliness and progress in technology. The development of combined cycle unit makes power generation, particularly using natural gas, more efficient. Blanco *et al.*(2006), for example, observed that combined cycle with natural gas was not only more efficient but saves 33% of CO_2 emission compared to conventional cycle. The latter based on natural gas saves up 29% of CO_2 emission compared to when it is based on fuel oil (however *cf*:Bolattürk, 2006; Lazzarin and Noro, 2006; Sonibare and Akeredolu, 2004).

In 2007 only about 1.2 BCF was used per day for power generation and this has been predicted to increase to 6.2 BCF per day by 2011 in order to hit the target additional 14.7 GW (Ashby and Ahemba, 2008). Ibitoye and Adenikinju (2007) analysed the future electricity requirement in Nigeria and conclude that for the nation to fairly rapidly achieve the millennium development goals and become an industrialising nation, it is necessary for the current electricity supply to be increased from the current 6500 MW to over 160 GW by 2030. This requires utilising vast amount of energy source over 50% of which has been predicted to be natural gas. Considering the limited hydropower resource (14,750 MW) of the country, meeting the 160 GW requirements will require using over 30 BCF of natural gas.

In transportation sector, 98 % of fuels used are oil-based (Isenberg, 1999). This is however gradually changing as the use of natural gas vehicles (NGVs) is on the increase. Currently there are over 130, 000 NGVs based on compressed natural gas (CNG) in the US alone and over 5 million worldwide. In fact about 25% of all new transit buses orders in the US are NGVs. These vehicles and their engines in over 150 models are now manufactured by about 50 different manufacturers (NGVC, 2006; USDE-EERE, 2006a). It is therefore obvious that the future of automobiles belongs to NGVs despite apparent shortcomings.

The main problems associated with NGVs are: they have relatively low mileage, low payload capacity and high cost. These are in addition to requirement for replacement or modification of existing oil-based infrastructure. However, when other factors are taken into consideration, NGVs are generally attractive. NGVs have lower emissions, favourable cost, much longer (2-3) service years, availability and comparable acceleration, horsepower and cruise speed (USDE-EERE, 2006a). Moreover, US fuel price report of June 2006 shows that CNG was cheaper than gasoline by as much as 94 cents on energy basis (USDE-EERE, 2006b). According to TIAX LLC projections, by 2010 NGVs will be highly competitive against diesel-based vehicles with high cost of NGVs engines and pump systems being upset by lower emission control system and fuel cost over lifecycle and more so NGVs will have significant advantage in lifecycle cost even when the crude oil price is at $60/bbl at 2005$ (Schubert and Fable, 2005). Furthermore, the increasingly stringent environmental policies are also in favour of NGVs.

Recently, the European Union and European Automobile Manufacturers Association (ACEA) signed an agreement in which by 2008 CO_2 emissions are to be cut down by as much as 25% from 1995 level of 186 g/km and even more in the subsequent years (Holman, 2001). These targets are quite easily achievable with NGVs. For example, in the US where 80% of air pollution in cities is caused by cars and trucks, it has been shown that use of NGVs can reduce CO emission by 90 – 97%; CO_2 by 25%; NOx by 25 – 60% and non-methane hydrocarbons by 50 – 75% with much fewer toxic and carcinogenic emissions (NaturalGas, 2004b).

Though this might not be a nationwide problem in Nigeria at the moment, it is in the major cities where poor evening visibility due smog is a common occurrence. The situation will certainly worsen with increasing population coupled with the general psychology of the people of correlating ownership of one or more cars with opulence. So, more cars (vast majority of which are second-hand) mean more emissions. This coupled with more industrial emissions will force the government to develop policies that will ultimately be favour of NGVs.

4. Natural Gas as Source of Liquid Fuel

What probably makes natural gas more versatile is the fact that technologies have been successfully developed to convert it to conventional fuels with added advantage of being less harmful to the environment. In addition, the natural gas conversion processes are environmentally friendlier than the conventional oil refining operations. Higher hydrocarbons and other fuel gases can be produced from natural gas often indirectly via synthesis gas (syngas), though direct methods are also being developed. Specifically, natural gas or methane is converted to syngas through steam reforming process etc depending on economic considerations and the intended use of the syngas (Vosloon, 2001). The syngas is then further processed into various other products through a number of novel industrial processes.

The indirect processes for conversion of natural gas include gas-to-gas (GTG) and gas-to-liquid (GTL) processes all of which involve conversion of the natural gas into syngas as intermediate step. GTG process is most importantly used to convert syngas to dimethyl ether (DME) via methanol. DME is a gas of great potentials as fuel (see Arcoumanis *et al.*, 2008) for comprehensive review). It can be used as substitute for LPG in diesel engines, thermoelectric power plants and fuel cells with almost no particulates or SOx emissions (Arcoumanis *et al.*, 2008; IDA, 2006; JDF, 2004).

GTL processes, on the other hand, produce liquid fuels through conversion of syngas to hydrocarbon liquids using Fischer-Tropsch synthesis. This technology has been in operation since 1950s as Sasol project in South Africa (Dry, 1990) and more recently in Malaysia (Sie *et al.*, 1991) for production of synthetic fuel, diesel, and other liquid fuels using syngas from coal and natural gas respectively. The technology gives high quality fuel and in greater quantity than refinery process (Dupain *et al.*, 2006; Sousa-Aguiar *et al.*, 2005). Two GTL processes specifically tailored to produce fuels with high octane ad cetane numbers are the methanol-to-gasoline (MTG) process and Mobil Olefins to Gasoline and Distillate (MOGD) process, respectively. MTG process uses ZSM-5 catalysts to convert methanol, from syngas, to branched alkanes and aromatic hydrocarbons in the gasoline range with high octane number (MacDougall, 1991). The MOGD process, on the other hand, uses special shape-selective catalysts to oligomerise light olefins into linear and slightly branched alkanes suitable for high cetane number diesel (Blauwhoff *et al.*, 1999). The olefins may be sourced from Methanol-to-Olefins (MTO) process (see below).

It is however worth noting that the main drawback to these technologies is, they are expensive compared to oil refining. This is mainly due to the fact that syngas production is capital inten sive process and account for about 60% of the CAPEX in GTL process. Hence, efforts are being geared towards developing catalysts for direct conversion of natural gas into liquid hydrocarbons thereby bypassing the syngas generation (Sousa-Aguiar *et al.*, 2005).

5. Natural Gas as Raw Material for Base Chemicals

Wet natural gas (associated and non-associated) often contains economic quantities of light C_{2+} paraffin gases. These are extracted, fractionated and steam cracked to ethylene, propylene and butenes and dibutenes, which are important base chemicals for the chemical industry (Moulijn *et al.*, 2001). These components are important in Nigerian context as about half of the nation's gas reserves of over 5 TCM is associated (always wet) (Sonibare and Akeredolu, 2006). Furthermore, technologies have been developed to achieve conversion, albeit indirectly, of the fairly inert methane into various base chemicals and intermediates used as raw materials for fine and speciality chemicals production (Moulijn *et al.*, 2001).

About 85% of chemicals are produced from few base chemicals which in turn are obtained mainly from crude oil, coal and natural gas. These base chemicals, which include syngas (from which ammonia and methanol could be produced), light olefins and aromatics, are used, through a number of conversion steps, to produce such consumer products as plastics, synthetic fibres, fertilizers, pesticides, drugs/pharmaceuticals, detergents, etc. (Moulijn *et al.*, 2001).

The base chemicals are traditionally produced from petroleum fractions/products. Lower *n*-alkenes (ethene, propene, butenes and butadiene) are mainly produced from steam cracking of ethane (mainly in the Europe) and naphtha (mainly in the US) while the aromatics (benzene, toluene and xylenes) are mainly produced from catalytic reforming of naphtha (Moulijn *et al.*, 2001; Speight, 1999). Increasing demand of liquid petroleum products, particularly gasoline and therefore naphtha, means increasing cost of producing base chemicals. Thus, where the feedstocks are cheaply available, syngas can be produced from gasification of coal, partial oxidation of fuel oil, otherwise it is mainly produced through steam reforming of natural gas.

Fortunately, the base chemicals can be produced directly or indirectly from natural gas. We have seen natural gas as raw material for syngas, and from it many fuels including liquid fuels are produced. These novel processes can be taken advantage of for production of light olefins and aromatics for the chemical industry.

Figure 4: Relative Distributions of Naphtha Cracker and MTO Process

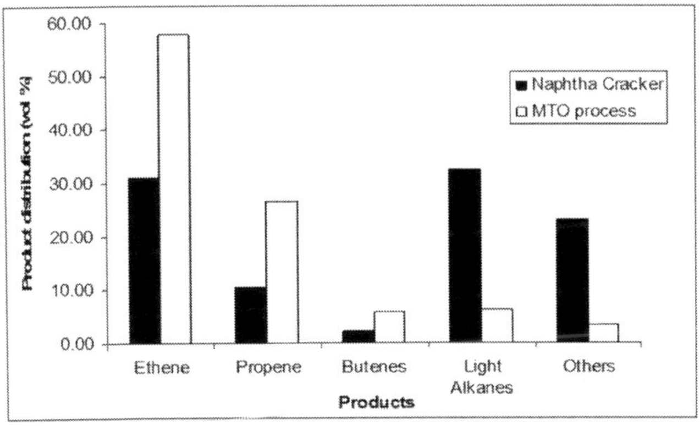

Source:Moulijn et al.(2001).

MTO process is a technology designed to produce light olefins from methanol obtained from syngas. The process uses novel shape selective catalysts (ZSM-34, SAPO-34 or Ni-SAPO) to literally terminate MTG reactions at olefins (Inui *et al.*, 1990; Inui and Takegami, 1982). The MTO products compare favourably to that from naphtha cracker (Figure 4) – an important oil refinery unit for sourcing light olefins (Moulijn *et al.*, 2001). Optimisation and commercialisation of this technology has been achieved by the Chinese Academy of Sciences' (CAS) Dalian Institute of Chemical Physics. The Dalian MTO process pilot plant has realized a 100% conversion rate of methanol with over 90% selectivity for ethylene and propylene (Liu *et al.*, 2000). Following this success the Institute signed agreement with Luoyang Petrochemical Engineering Co. Ltd. to build a 10,000 tonnes/year MTO commercialisation unit (CCR, 2005). Similarly, UOP LLC and Total

Petrochemicals signed joint agreement to build an integrated demonstration plant in Feluy, Belgium. The plant, started in autumn of 2008, consists of a UOP/HYDRO MTO process unit and a Total Petrochemical/UOP Olefin Cracking process unit, for cracking heavier olefins from the MTO unit (Total, 2010). The plant provides feed for large-scale polymerisation pilot plant (UOP, 2005; and see Keil, 1999; Stöcker, 1999;for detailed review). Furthermore, it has been shown by Dupain *et al.*(2006) that both liquid fuels and olefins (over 30%) can be produced simultaneously from Fischer-Tropsch waxes by doping the equilibrium catalysts with ZSM-5.

Light aromatics, particularly benzene, toluene and xylenes (BTX), on the other hand, can be sourced from MTG process which produces significant amount of these compounds in the gasoline range (see above). These important base chemicals can also be produced by conversion of MOGD, MTG and Fischer-Tropsch products using conventional catalytic cracker and catalytic reformer.

Considering the nation's vast amount of coal deposits at Enugu, natural gas-coal co-reforming process may also be adopted for syngas production.

6. Conclusion

Sustainability of the environment is probably the single most important contemporary discourse. It is widely acknowledged that anthropogenic activities, particularly use of fossil fuels, are detrimental to the environment. Such activities have been shown to generate the so-called greenhouse gases which contribute to global warming (Harrison, 2001a). Hence colossal efforts and resources are now directed towards using the environment in a sustainable way. Natural gas utilisation is one of most promising sources of energy that can help secure the environment in a sustainable way.

The above discussion clearly shows the future of fuel sourcing as that of utilisation and conversion of natural gas into various high quality and environmentally friendly fuels and raw materials for the chemical industry. This future will be dictated not only by producers of natural gas but perhaps more importantly by the custodians of the relevant conversion technologies. North America and Europe harbour over 40% of the world's natural gas proved recoverable reserves (Africa 7.5% only) (Aptroot, 2004). These same regions, and Asia (*ca* 11%), are leading in R&D in development of natural gas conversion technologies. This means both prices of raw material (natural gas) and products (synpetroleum and base chemicals) will be controlled by developed countries and thus developing countries like Nigeria will only be suppliers of raw materials and consumers of products with much less bargaining power than is currently obtainable with respect to petroleum. It is therefore imperative that countries like Nigeria be actively involved in utilisation of these technologies.

Furthermore, production and industrialisation of consumer products stimulate development of skills and technology through constant generation of problems that need to be solved which more often than not require involvement of the academia. This thus leads to industry-academia cooperation resulting in synergy in development. The field of natural gas utilisation and processing is a very challenging one and a fertile area where such interaction will foster cooperation with a resultant consequence in industrial development of the country.

References

Arcoumanis, C., Bae, C., Crookes, R. and Kinoshita, E. (2008), 'The potential of di-methyl ether (DME) as an alternative fuel for compression-ignition engines: A review', in *Fuel*, 87(7): 1014-1030.

Ashby, T. and Ahemba, T. (2008), 'Nigeria approves new rules for natural gas', In: *Reuters, UK - Business and Finance*.

Blanco, J.M., Mendı´a, F. and Peña, F. (2006), 'Comparative analysis of CO_2 and SO_2 emissions between combined and conventional cycles with natural gas and fuel oil consumption over the Spanish thermal power plants', in *Fuel*, 85: 1280 - 1285.

Blauwhoff, P.M.M., Gosselink, J.W., Kieffer, E.P., Sie, S.T. and Stork, W.H.J. (1999), 'Zeolites as Catalysts in industrial processes', in Weitkamp, J. and Puppe, L. (eds.), *Catalysis and Zeolites: Fundamentals and Applications*, Berlin: Springer-Verlag.

Bolattürk, A. (2006), 'Determination of optimum insulation thickness for building walls with respect to various fuels and climate zones in Turkey', in *Applied Thermal Engineering*, 26: 1301 – 1309.

CCR, (2005), Methanol-to-olefin production process.*China Chemical Reporter*, *16*: 15. Available at: http://www.ccr.com.cn/new/search_list.aspx?key=Methanol-to-olefin+production+process. Accessed on 17 September, 2006.

Cook, L. (2006), *Fuelling the future – the transition to a global LNG market*, London: World Energy Council.

Copley, C. (2004), 'Coal (including Lignite)', *In:*2004*Survey of Energy Resources*. Elsevier, Amsterdam, pp. 464. Available at: http://www.worldenergy.org/publications/324.asp. Accessed on 15 July, 2007.

CP&S, (2009), 'Annual Statistical Bulletin', Nigerian National Petroleum Corporation, Abuja, p. 28.

Dry, M.E. (1990), 'The Fischer-Tropsch process – commercial aspects', in *Catalysis Today*, 6: 183-206,

Dupain, X., Krul, R.A., Schaverien, C.J., Makkee, M. and Moulijn, J.A. (2006), 'Production of clean transportation fuels and lower olefins from Fischer-Tropsch Synthesis waxes under fluid catalytic cracking conditions: the potential of highly paraffinic feedstocks for FCC', *Applied Catalysis B: Environmental*, 63: 277-295.

Harks, E. (2006), 'Supply and Demand: International interests and tensions', In: *World Energy in 2006*, pp. 12 - 15, World Energy Council, London.

Harrison, R.M. (2001a), 'Air pollution: sources, concentrations and measurements', in Harrison, R. M. (ed.), *Pollution: Causes, Effects and Control*, Cambridge, UK: Royal Society of Chemistry.

Harrison, R.M. (2001b), 'Chemistry and climate change in the troposphere', in Harrison, R.M. (ed.), *Pollution: Causes, Effects and Control*, Cambridge, UK: Royal Society of Chemistry.

Holman, C. (2001), 'Control of pollutant emissions from road traffic', in Harrison, R. M. (ed.), *Pollution: Causes, Effects and Contol*,Cambridge, UK: Royal Society of Chemistry.

Ibitoye, F.I. and Adenikinju, A. (2007), 'Future demand for electricity in Nigeria', in *Applied Energy*, 84: 492 - 504.

IDA, (2006), About DME. Available at: http://aboutdme.org/index.asp?sid=48. Accessed on 17th September, 2006.

Inui, T., Phatanasri, S. and Matsuda, H. (1990), 'Highly selective synthesis of ethene from methanol on a novel nickel-silicoaluminophosphate catalyst', in *Journal of Chemical Society: Chemical Communication*, 205-206.

Inui, T., Takegami, T. (1982), 'Olefins from methanol by modified zeolites', in *Hydrocarbon Processing*, 6: 117-120.

Isenberg, G. (1999), 'Assessment of automotive fuels', in *Journal of Power Sources*, 84: 214-217.

JDF, (2004), 'DME as a Fuel'. Available at: http://www.dmeforum.jp/index_e.html. Accessed on 17th September, 2006

Keil, F.J. (1999), 'Methanol-to-hydrocarbons: process technology', in *Microporous and Mesoporous Materials*, 29: 49-66.

Lazzarin, R. and Noro, M. (2006), 'Local or district heating by natural gas: which is better from energetic, environmental and economic point of views?', in *Applied Thermal Engineering*, 26: 244-250.

Liu, Z., Sun, C., Wang, G., Wang, Q. and Cai, G. (2000), 'New progress in R&D of lower olefin synthesis', in *Fuel Processing Technology*, 62: 161-172.

MacDougall, L.V. (1991), 'Methanol to fuels routes - the achievements and remaining problems', in *Catalysis Today*, 8: 337-369.

Moulijn, J.A., Makkee, M. and Van Diepen, A. (2001), '*Chemical Process Technology*', Chichester: John Wiley & Sons Ltd.

NaturalGas, (2004a), 'Electric generation using natural gas'. Available at: http://www.naturalgas.org/overview/uses_eletrical.asp. Accessed on: 28 August, 2006.

NaturalGas, (2004b), 'Pollution from the transportation sector - natural gas vehicles'. Available at: http://www.naturalgas.org/environment/naturalgas.asp#pollution. Accessed on: 27 August, 2006.

NaturalGas, (2004c), 'Residential uses'. Available at: http://www.naturalgas.org/overview/uses_residential.asp. Accessed on: 28 August, 2006.

NGVC, (2006), 'About natural gas vehicles'. Available at: http://www.ngvc.org/ngv/ngvc.nsf. Accessed on: 27 August, 2006.

Schubert, R.K. and Fable, S. (2005), 'Comparative costs of 2010 heavy-duty diesel and natural gas technologies final report', TIAX LLC. Available at: http://www.eere.energy.gov. Accessed on: 28 August, 2006.

Sie, S.T., Senden, M.M.G. and Van Wechem, H.M.H. (1991), 'Conversion of natural gas to transportation fuels via the shell middle distillate synthesis process (SMDS)', in *Catalysis Today*, 8: 371-394.

Sonibare, J.A. and Akeredolu, F.A. (2006), 'Natural gas domestic market development for total elimination of routine flares in Nigeria's upstream petroleum operations', in *Energy Policy*, 34: 743-753.

Sonibare, J.A. and Akeredolu, F.A. (2004), 'A theoretical prediction of non-methane gaseous emissions from natural gas combustion', in *Energy Policy*, 32: 1653-1665.

Sousa-Aguiar, E.F., Appel, L.G. and Mota, C. (2005), 'Natural gas chemical transformations: the path to refining in the future', in *Catalysis Today*, 101: 3-7.

Speight, J.G. (1999), *'The chemistry and technology of petroleum'*, CRC, New York, p. 934.

Stöcker, M. (1999), 'Methanol-to-hydrocarbons: catalytic materials and their behavior', in *Microporous and Mesoporous Materials*, 29: 43-48.

Total, (2010), 'Research - total petrochemicals research feluy'. Available at: http://www.be.total.com/content/en/research/Research-Total-Petrochemicals-Research-Feluy.aspx?lg=en. Accessed on: 28 August, 2010.

UOP, (2005), 'UOP LLC press release'. Available at: http://www.uop.com. Accessed on: 10 September, 2006.

USDE-EERE, (2006a), 'Clean cities alternative fuel price report June 2006'. Available at: http://www.eere.energy.gov. Accessed on: 28 August, 2006.

USDE-EERE, (2006b), 'Natural gas'. Available at: http://www.eere.energy.gov/afdc/altfuel/natural_gas.html. Accessed on: 28 August, 2006.

Vosloon, A.C. (2001), 'Fischer–Tropsch: a futuristic view', in *Fuel Processing Technology*, 71: 149-155.

Aptroot, (2004), 'Natural gas', *In:*2004*Survey of Energy Resources*. Elsevier, Amsterdam, pp. 464. Available at: http://www.worldenergy.org/publications/324.asp. Accessed on 15 July, 2007.

Weissermel, K. and Arpe, H.-J. (2003), *Industrial organic chemistry*, Weinheim: Wiley-VCH. 511 pp.

Chapter 8

Uyinmwen and the Development of an Indigenous Genre in the Edo-Benin Home Video Industry, Nigeria

Uyilawa Usuanlele

1. Introduction

"...the Nigerian nation is simply too big and diverse to be fully represented by any one kind of film." (Haynes and Okome, 1998)

The Nigerian movie industry which has been gradually developing in big screen celluloid formats and largely as a reaction to Western stereotyping of Africa) in the early 1970s and 1980s (Thackway, 2003 and Ofeimun, 2004) suddenly became a big industry of world acclaim known as Nollywood in the 1990s. This was the outcome of a change to themes dealing with indigenous African cultural values and the adoption of the video format for home use given the popularity of Video Cassette Recorders in Nigeria since the late 1970s. The change was also characterized by production that appealed to an indigenous constituency and adaptation to their yearnings (Ofeimun, 2004), hence the production of home videos in various Nigerian indigenous languages. The economic undercurrent for the change to the home video format has also been noted by Haynes to be borne out of economic crisis of the 1980s (Haynes, 1995). But this argument is largely based on the experience of Lagos and the Yoruba folk cinema without any reference to developments in other parts of Nigeria.

This argument is traceable to the fact that documentation of the history of film development in Nigeria has largely been restricted to works in English language and the three largest language groups of Nigeria namely Hausa, Yoruba and Igbo. The contributions and development of films in the less populous languages are either treated as footnotes (Haynes and Okome, 1998) or in some cases denied as attested by Oluyinka Esan's 2008 article in *Participations: Journal of Audience and Participation Studies,* which stated that:

> Nollywood evolved around a dominant language structure which also reflects the original geopolitical structure of Nigeria. The industry thus has 3 distinct sectors; Yoruba movies, Igbo movies and Hausa movies. The movies tend to reflect the cultures in these areas and are often produced and distributed from the commercial nerve centres in these regions and Lagos - the nation's commercial capital (Esan, 2008).

This is in spite of the production of home videos in the languages of the smaller groups like *'Benin-Edo'*, *'Esan-Edo'*, *'Etsako-Edo'*, *'Urhobo'* and *'Itsekiri'* in the former Midwest area (Usuanlele, 1999a) and *'Igala'* and *'Ebira'* in Northern Nigeria. It is also a known fact that actors and in some cases scripts from these small language areas have been greatly utilized in the production of films in the more populous languages. Artistes like Sam Loco Efe (Edo), Don Pedro Obaseki (Edo), Ejiro (Urhobo), the Amata family (Isoko) Amenze Imarhiagbe (Edo) Mercy Aigbe (Edo) are known pillars of the industry in both English, Igbo and Yoruba language movies. Such denial and neglect deprives the world of the knowledge and enrichment which the smaller languages contribute and will continue to contribute to bettering the industry and humanity.

This chapter draws attention to the home video industry among one of the small language or minority language groups specifically the Benin-Edo people of Southern Nigeria. It looks at the development of the home video film industry among the Edo people and their contributions to pioneering the industry in Nigeria. It also discusses the seemingly new indigenous *Uyinmwen* type genre, its format, objective and trends in its adoption among Edo home video producers as well as its potentials for the future of the industry.

2. Development of Home Video Film Industry in Benin City

The development of Nollywood home video industry in Lagos and Yoruba speaking area of Nigeria can be attributed to the cultural and economic experiences of the area. This film industry was the handiwork of the professional filmmaker and later the grafting of the traditional Yoruba travelling theatre on the film medium. It soon became a huge success with the involvement of other non-Yoruba professionals and entrepreneurs. The experience of Benin City and its Edo people is not too dissimilar except for their different trajectories. This section contributes to understanding the development of the Nigerian home video industry by describing the Benin City and Edo experience which virtually bypassed the big cinema straight to home video format, for differing reasons. It also shows that the Benin City experience predated the Nollywood and Yoruba adoption of the home video format.

Firstly, the Edo people and their Benin City capital were not privileged with entrepreneurs to invest or venture into the big screen[1] as happened among the Yoruba in the late 1970s. However, the local popular/folk theatre which had developed since the colonial period in Benin City was augmented by the establishment of the Midwest Television in 1973. This provided another outlet for artistic production and access to a larger audience. But more importantly, it brought with it the realization that their production can be documented and archived in other formats for entertainment purposes at the convenience of the audience. A similar development took place in Benin in the early 1960s when the British Council was established and provided technological gadgets to promote the Arts. The late Ikponmwonsa Osemwegie, a pioneer folk dramatist, poet and playwright got a tape recorder from the Council and employed it to record the performances of his Edo language plays and poems, which he played to audiences in the city and private homes of prominent citizens/patrons (Osemwegie, 1997). But in later years, some phonodisc producers like Akpola Records, Omada Records and Supreme phonodisc recorded the dramatic works and folk stories of some of the theatre groups and mass produced them in vinyl for the local market.(Usuanlele, 1998). This mode of production lost popularity with the advent of television and died with the introduction of a new technological device, the Video Cassette Recorder.

The entry of the Video Cassette Recorders (VCR) into the Nigerian market in the late 1970s and early 1980s changed the culture industry and the fortunes of local stage artistes in Benin City. The device was acquired by the rich for their home use, but the entertainment choices available to them were only foreign video films. The few local big screen films in English and Yoruba did not exist in the video format. Realizing the huge market potential for local big screen films, the stage and television plays of local theatre and cultural groups, some production staff of Nigerian Television Authority and Bendel State Television, Benin City stepped in to meet the market demand. They, albeit, illegally dubbed and sold VHS recordings of stage and studio productions on demand.[2] The local folk theatre practitioners bought the idea as they recorded their performances and sold the videos. This started the process of production and sale of video films of local folk productions in Benin by 1982. Some examples of the stage drama performances that were recorded and sold in video format in this early period are *Okiekie aren evbayakon* by Inneh Cultural group, *Oyi nere egbo atan*, and *Erhitio Sole Sole* by Ozin

Oziegberuwa led Otu Uru Edo cultural group.

The reproduction of stage drama in Edo language by the folk theatres in VHS format and sale in Benin City area coincided with the economic crisis of the early 1980s which brought about increased social insecurity. This affected the Cinema houses as a major leisure and entertainment outlet. Compounding the situation further were the indigenization decree of 1980 and the foreign exchange crunch which constrained the dominant Lebanese cinema owners to only the importation of poorly rated Indian and Chinese movies and outdated Westerns with resultant poor patronage. Competition from the expanding television networks with twenty four-hour weekend transmissions of both foreign films and local stage plays further eroded the cinema houses patronage. The cinema houses consequently closed down in Benin City in the early 1980s.

These developments changed the fortunes of local folk theatre groups and encouraged them into more production of home videos of their stage and studio performances. But it also led to a gradual shift from studio and stage productions to outdoor locations. This transition was pioneered by Ms Aghabiomo Ogbeiwi led Uyi-Edo theatre troupe which released *Uvbi Udefiagbon ne Ovbiosa* in 1988 (Personal Communication with Aghabiomo Ogbeiwi, 2010). Though she started as a folk theatre artist, she later had some theater education at the University of Ife and worked with the Nigerian television Authority before transferring to Bendel Television Authority, Benin as an Edo language announcer, newscaster and artist. Her exposure to the big screen Yoruba folk cinema in Ife in the early 1980s inspired her to tap her television connections to shoot the first Edo home video on location with her Uyi-Edo Theatre Troupe.

This experiment of video film production in local Edo language solely for the market suffered an initial problem of marketing as distribution outlets were practically non-existent. The only means of distribution was the theatre groupswhich made copies from master tape and sold it to members of the public on demand. Lack of capital or financiers to advertise and promote the videos made the videos unknown to the wider public and affected demand. This resulted in poor returns on investment entailing the cost of transporting, housing and feeding artiste and crew at locations. The situation was further compounded by the problem of piracy, the unauthorized reproduction and sale of works under copyright. Taken together, the poor marketing network and returns on investment as well as piracy discouraged further production from Uyi-Edo and other folk theatre groups.

The problems of marketing and piracy were later solved with the rise of Nollywood in Lagos in the early 1990s. The music/video entrepreneurs of Lagos and Onitsha employed their national networks to create the mass market. Their counterparts in Benin not only marketed the Lagos Nollywood videos, but started to take an interest in video production in the local Edo language as well. This development created the much needed environment for the revival and further evolution of the Edo folk home video Industry. The local Edo folk theatre groups cashed in on this opening and commenced production of home videos for the mass market.

Ozin Oziengberiuwa a. k. a. Erhitiro as leader of (Otu Urhu Edo theatre troupe) who had been involved in stage and television production and dubbing of home video of their plays on demand (Usuanlele, 1998) quickly exploited this new niche and produced and released - *Ewemade* in 1997 (Usuanlele,1998). Since he was not obliged to pay his theatre troupe members, his major financial obligations were to the hired technical crew and transportation services. With the production of *Ewemade* completed, Erhitio promoted the video film with radio and television advertisements and reviews in newspapers that combined to ensure its marketing success. Despite this, he suffered losses due to piracy. Notwithstanding, he was resolute about producing more videos and this made him to henceforth work with pirates to produce and market his films. This opened the floodgate to further production of home videos in Edo language and serves as the revival of the almost defunct Edo home video industry in Benin City. It was totally devoid of any financial and intellectual investments by businessmen and scholars respectively, which was the case in the 1980s with the Yoruba and Igbo home video.

Following the success of *Ewemade,* the former music pirates and some university trained theatre artists started venturing into Edo home video industry as entrepreneurs and artists respectively. It became so lucrative that entrepreneurs started imitating their Lagos and Onitsha counterparts by buying script and employing directors and artists to produce the scripts for them. In one case, Mr. Monday Osagie, a. k. a. Soul 2 Soul a former music pirate turned entrepreneur acts in the home videos produced by his video production company. The English language drama practitioners introduced the themes relating to cults, gangsters, Pentecostal Christianity, witchcraft, and comedic genres of Lagos into the Benin-Edo industry. This started turning the Edo home videos into poor imitations of the Lagos variant. By 2000, there was an average of fifteen to twenty home videos produced in Benin City every month (Personal Communication with Henry Legemah, 2010). This development brought about more competition and relative improvement in quality of movies. But there was no corresponding increase in editing studios in Benin City; consequently, some of the producers had to take their works to Lagos and Onitsha for editing. There developed a glut of the market with Edo videos. This soon forced the producers and artists to establish a regulatory marketing body to control and decide the number of videos to be released monthly. This helped reduce the number of films released monthly to an average of eight to ten in late 2000s.[3]

3. Uyinmwenand the Emergence of a New Indigenous Genre

The Edo home video industry had developed on the basis of self efforts of the local folk theatre groups in Benin City who relied on their indigenous knowledge. The movies were initially based on folk history and lores which Benin is very rich in. The story line is usually straight and the actors act in a traditional palace or village setting, the characters don traditional costumes and employ music and dance where required along with normal dialogue. The entry of the university trained theatre practitioners introduced themes dealing with cults, gangsters, prostitution, witchcraft, and Pentecostal Christianity to the industry divested it of its Edo cultural stamp, with the exception of the language. It was more of an imitation of Lagos Nollywood videos except for its poorer quality.

The debut of Erhitiro Sole Sole and Enofe Igbinomwanhia

produced *Uyinmwen* home video in 2008 seems to signify a drastic change in the Edo home video genre- different in conception and production format. *Uyinmwen* employs the total theatre concept and explores the whole repertoire of Edo orature and musical experience to achieve its objectives of entertainment and teaching of morals to youths and adults who seem to be increasingly adopting Western values and modernity. This genre opens with a dramatized straight dialogue about an issue before unfolding the storyline and message through a combination of storytelling, call and response, praise poetry, choral music, and drama rolled into one package. It is unabashedly didactic.

The first film to use this genre is *Uyinmwen* (meaning Character) and revolves around the issues of child upbringing, formation of character and its consequences for adult life. It opens with the disobedient act of a modern teenage girl who wants a life on the fast lane and this earns her and her visiting boyfriends' corporal punishment from her Army General father. But these teenagers are rescued from punishment by the teenager's grandfather. The grandfather Erhitio then employs the story telling format accompanied with *Akpata* (harp) to teach the teenagers and their parents the traditional method of child upbringing. The embedded story is that of Iroguehi, a teenage girl who disobeys and disrespects her parents, does not conform to the expected social mores and values and fails to accept correction of her behaviour while growing up. She eventually marries and was found wanting in values and household chores, which unfortunately led to her being eventually evicted from matrimony in shame and tears. The story changes the heart of the General's teenage daughter who consequently apologizes and promises to change her behaviour. She also invites her grandfather to teach her schoolmates about the cultural values of the past. The movie ends with the teenage girl's school mates and teachers learning the various traditional dance forms from her grandfather. This didactic message of the story is backed up with the indigenous call and response poetry, proverbial duels and wisecracks and various forms of Edo music, dance sketches as well as grandfather's narration interspersing it. All of these combine to make this genre a totally new total theatre experience.

An evaluation of this *Uyinmwen* genre with the genre of the

Yoruba folk theatre and video shows them to be different. According to Adelugba and Obafemi, the Yoruba folk theatre:

> Performance consists of an opening glee, a spectacular parade of dance and songs by chorus dancers led by a lead singer, who ultimately is the troupe's leader, followed by a more elaborate dance, often indistinguishable from the opening glee, and lastly the drama. The dramaturgy assimilated the Alarinjo masque festival form and anticipates the total theatre aesthetics of the future drama of Nigeria. It consists of the dynamic interplay of the visual, verbal, and kinetic elements of music, dance, poetry and songs, with invited audience participation, as in the traditional festival theatre (Adelugba and Obafemi, 2004).

Jonathan Haynes also observed similar trend in the Yoruba folk cinema and video:

> A principal genre of the Yoruba films is the costume drama, which aims to recapture the vanished splendor of traditional Yoruba culture in all its metaphysical, social, and aesthetic integrity. It relies heavily on dancing and music and recreated festivals to do so; the visual aspect of these films can be stunning. Both the theme and the multi-media treatment are extensions of the Travelling Theater, as pioneered by Ogunde (Haynes, 1995).

In spite of what appears to be a similarity with the Yoruba theatrical genealogies as sketched in the foregoing, there does not seem to be a Yoruba influence on the Uyinmwen genre of home video of the Edo. The inspiration for the *Uyinmwen* type genre is derived from the Edo folkloric past. It weaves materials from the whole repertoire of Edo orature and musical traditions including the colonial inspired popular music and dances to dramatize the story and its message. The call and response dialogic poetry is reminiscent of moonlight game in which the acting King or Queen is allowed to freely tease and insult the subjects to be chosen for domestic service known as *Izogie*. Another folkloric material is the *Egbo*poetry which Benin Queens use to communicate their heart desires to the King during *Ibota* –family evening entertainment (Erhahon, 1998). In addition, there is the *Itan vbe Ivbe* or proverb duel in which dialogic barbs are exchanged through appropriately chosen proverbs, another format for communicating in the traditional harems. To these are added, various dance forms and appropriate choral music to drive home the message of the story and to entertain. The costume is not always and necessarily traditional as this is influenced by the period and setting of the story. Another major characteristic that is discernible from this new genre is that the stories are usually short and in some cases sketchy.

The beginning of the *Uyinmwen* genre is traceable to the collaboration between modern popular Edo musicians and folk theatre practitioners in the production of promotion videos. With the rise of the folk home video cinema, the popular musicians commissioned folk theatre artistes to create informercials or short films to advertise and promote their new albums. The artistes, in turn, use the folkloric resources and genre for producing such informercials. Another factor that further helped the birthing of Uyinmwen genre is the fact that some of the folk theatre artistes are also the popular musicians in their own right. Examples are Enofe Igbinomwanhia (a.k.a. Akpaka), Wilson Ehigiator (a.k.a. Akobe Eghian), and late Erharuyi Ogbeide (a.k.a. Ukeke) and so on. They were not limited only to music and folk theatre, they also engaged in other areas of the entertainment industry as well. What is noteworthy, however, is that the popularity and success of the informercials contributed to the development of the *Uyinmwen* genre namely the infusion of indigenous music with poetry.

Making the *Uyinmwen* genre, a form of Edo home video is largely the handiwork of Ozin Oziengberuwa a.k.a. Erhitiro (with the input of his former apprentice) Nofe Igbinomwanhia (a.k.a. Akpaka99), a younger cultural activitist, filmmaker, actor and musician trained in the indigenous folk theatre tradition. This was made possible by Erhitio's long dedication to preservation and promotion of Edo culture. Erhitiro Sole Sole had his primary education in Ugbogiobo village, and then trained as an artisan (welder) in Benin City. He ventured into indigenous theatre and cultural performances in the early 1960s when he joined Amazeevbomwan theatre troupe of Benin formed by pioneer indigenous playright and poet Ikponmwonsa Osemwingie. He later abandoned his artisan trade to found and lead his own *Otu Uru Edo* (Voice of Benin) cultural and drama group, which have always included his wives and children. He wrote and improvised his own plays and songs and even created a musical instrument made from calabash and some other materials that sound like a trumpet. In the 1970s and 1980s, he was featuring on NTA Benin programmes on some of his indigenous plays where he earned the stage name *Erhitiro sole sole.*

In spite of the exposure to the modern television and video format, Erhitio has maintained his obsession with the indigenous culture and genre and remained committed to a neo-traditionalist ideology in his plays and films. (Usuanlele, 1998) His movie *Uyinmwen* is a clarion call to uphold and use traditional Edo virtues of morality, respect for elders, appropriate greeting code/salutation, work and culture. He returned to this neo-traditionalist theme in *Are Wiri* (heritage is lost) where the emphasis is the value of teaching Edo language to children and the consequences of losing the mother tongue and indigenous etiquettes and code of conducts. Because of the sketchy nature of the stories, he adds segments of lost and dying dance forms to entertain and teach the younger generation.

As already noted, more films in the template of *Uyinmwen* genre have been produced and released. These video films were produced after Erhitiro and Nofe Igbinomwanhia's *Uyinmwen* huge market success. Even some musical production of the 1990s like Chief Ewaensetin Eholor's *Emwen Evbo ebo ne o re na* have been made into a film in that genre. Chief Ewaensetin Eholor of Igieduma, another notable folklorist whose artistry goes back to the 1950s in his rural Igieduma/Ehor area is rather eclectic in his combination of various musical and poetic forms in his performances and this can be discerned from his debut album/cassette *Emwen Evbo ebo ne re na* in 1997 (Usuanlele, 1999b) *Emwen Evbo Ebo* combined virtually all the elements of story telling, drama, dialogues and singing, excepting the, *Egbo* poetry and the *Itan vbe Ivbe* proverb duel.

The *Uyinmwen* genre has increasingly been adopted by other video film makers and producers. Entrepreneurs have commissioned renowned cultural and theatre troupes and individual artists to write and produce films in this *Uyinmwen* genre. Some of the films are *Ukhuedo* (meaning the heritage of Edo) parts 1 (2009) and 2 (2010), *Iba noreze,* 2009 (meaning the misdeeds of the youths) *Are wiri* (meaning heritage is lost) 2010, *Omobe* (meaning Bad child) 2010, *Oguomwandia* (Servant) 2010, *Wagie mwin ma gha yeman* (let us value our heritage) 2010 and so on. All these movies are didactic and critical of the modernity embraced by Edo youth and women, who have no restrain in the adoption of western and foreign values that are deemed antithetical to Edo values and beliefs. Though the genre is largely neo-traditional in the vision of society it projects, it employs episodic approach or sketchy stories to put its message across. *Ukhuedo* Part 1 and 2 has four episodes each, while Eghe N'Ova's *Iba noreze* has three. *Iba noreze* is also eclectic as one sketch challenges the traditional practice of polygamy by Edo men which undermines women. Though the *Uyinmwen* genre has been adopted by other producers, it remains to be seen if it can be sustained and improved upon.

4. Conclusion

This chapter examined the development of the home video film industry among the less populous Benin-Edo from its stage and studio drama days of the 1950s to the emergence of the new genre used in *Uyimwen* and pioneered by Ozin Oziengberuwa and Nofe Igbinomwanhia in 2008. It debunks the view that the home video industry was the exclusive preserve of the majority Hausa, Yoruba and Igbo ethnicities and shows the contribution of the less populous ethnic groups. It also shows that the development of home video industry in Edo Language in Benin City since the 1980s was independent and was different from developments elsewhere in Nigeria. It described the emergence of *Uyinmwen* genre from the indigenous musical and theatre traditions of the Edo people. It is didactic and neo-traditionalist in nature. It concludes that the *Uyinmwen* genre might not remain neo-traditionalist particularly with its increasing adoption by different producers. The questioning of traditional polygyny in *Iba ne Ore ze* is a pointer to deviation from the neo-traditionalist ideology of its founders.

Notes

1 The only big screen film made was *Oba Ovonramwen N'Ogbaisi* funded by the Midwest State Government in 1974 and a few documentaries on Edo Kingship. See: http://www.cgore.dircon.co.uk/films/htm.

2 Personal knowledge and patronage of these pirated works from staff members of Television stations.

3 Interviews with Henry Iyobosa Legemah, Independent Film Producer, Benin City, July 2010, and Aghabiomo Ogbeiwi, Founder and Leader of Uyi-Edo Theatre Troupe International of Benin City, New Jersey, USA, August 2010.

References

Adelugba, D. and Obafemi, O. (2004), 'Anglophone West Africa: Nigeria', Martin Barham, *A history of theatre in Africa,* Cambridge: Cambridge University Press.

Erhahon, Ivie (1998), 'Egbo: A form of communication among the Edo' in *Benin Studies Newsletter,* 3(3): September.

Esan, O. (2008), 'Appreciating Nollywood: Audiences and Appreciation', *Participations,* Available at: http://www.participations.org/Volume 5/Issue1special 5_01_esan.htm).

Haynes, J. (1995), 'Nigerian Cinema: Structural Adjustment', in *Research in African Literature,* 26(3): 97-119.

Haynes, J. and Okome, O. (1998), 'Evolving Popular Media: Nigerian Video Films' in *Research in African Literatures,* 29(3): 106-128.

Ofeimun, O. (2004), 'In defense of the films we have made', in *West African Review,* Available at: http://www.africaknowledgeproject.org/index.php/war/article/view/307).

Osemwegie, I. (1997), 'Bleak future for Edo Literature, if…', in *Interview, inBenin Studies Newsletter,* 2(1 &2): 2-4.

Thackway, M. (2003), *Africa shoots back: Alternative perspectives in Sub-Saharan Francophone African Film,* Oxford: James Currey.

Usuanlele, U. (1998), 'A Patriotic voice of Edo lore', *The Guardian,* June 3.

Usuanlele, U. (1999a), 'Urhobo movies rise, Edo's in glut' *The Guardian,* September 16.

Usuanlele , U. (1999b), 'A breather for endangered Edo oral genre', *The Guardian,* April 7.

Chapter 9

Barriers towards the Development of Emerging Contractors in the Limpopo Province, South Africa

Lawrence K. Tshivhase & Zeleke Worku

1. Introduction

In many economies around the globe, construction industry plays a significant role in promoting economic growth, infrastructure development and job creation. In South Africa, the role of the construction industry in economic development and solving socio-economic challenges remain central. Because of the neglect of infrastructure development during apartheid, there remains a significant infrastructure opportunities across various sectors of construction.

The persistent nature of demand for the construction of major infrastructure projects in South Africa is the result of macro-economic matters such as global markets, political climate, return on investments and government collection of tax revenue (National Advisory Council on Innovation (NACI, 2003). Whilst government is creating opportunities in the construction industry, there are various constraints impacting on the performance of emerging contractors in Limpopo Province of South Africa. Due to the legacy of apartheid practices, which denied black-owned businesses to participate meaningful in the construction sector, the South African construction sector is currently dominated by the discrepancy between white and black-owned businesses. This has unfavourably affected black-owned businesses mainly attributed to historical deficiencies in racially skewed education system, income and economic empowerment (Berry *et al*, 2002: 8). According to Diale (2009), supply side constraints have a negative impact on government position to promote and support small businesses. Tshivhase (2007) established that more than half of new and emerging contractors in Polokwane in Limpopo Province of South Africa fail and go out of business for various business reasons. These comprised of challenges related to shortage of skills, administrative support and access to finance.

This study used the term "emerging" to refer broadly to growth towards maturity, reaching advance stage and the realization of full potential of emerging contractors in the Limpopo Province. The South African Construction Industry Development Board (CIDB) uses grading system as a method to measure the size and capacity of a contractor to perform a particular project (Refer to Table 1). Based on CIDB grades, the study considered contractors that are still in the development stage. In this regard, contractors holding CIDB grade of 2 to 7 were investigated. It is believed that contractors within these grades are true representative of emerging construction sector in Limpopo Province and are normally confronted by many challenges that threatened their existence and continued development within the construction industry. Furthermore, contractors holding CIDB grade of 8 to 9 were not investigated as seen to have progressed far beyond stage of being emerging with some of these contractors regarded as the global players in the construction industry.

2. Literature Review

The literature review gives a background concerning the reform of procurement for construction contracts in South Africa with an emphasis of promoting small and emerging contractors in the construction industry since the establishment of a democratic state. This also highlights the formation of CIDB aimed at regulating the construction sector as well as promoting emerging sector. As part of economic transformation, the review underscores the Construction Industry Sector Charter (CISC) that drives the empowerment of emerging contractors in the construction industry. Despite the introduction of these policies to create enabling environment for emerging contractors, challenges that were found to continue to affect emerging contractors in Polokwane in the Limpopo Province are also summarized.

Procurement for construction contracts in South Africa

Shortly after the establishment of democratic government in South Africa, public works reformed public sector procurement for construction projects (Watermeyer *et al.*, 1998). The reforms affirmed the transformed environment in South Africa associated with the achievement of government socio-economic objectives. Some of the plans which encompassed the reformation of procurement for construction contracts are as follows:

- Promoting easy access to tendering information;
- Withdrawal of sureties on construction contracts having a value of at least less than R 100 000;
- The unbundling of large projects into smaller construction projects;
- The promotion of early payment cycles by various spheres of government;
- The development of a preference system especially for small and medium enterprises owned by black people;
- The simplification of tender submission requirements;

- The classification of contracts according to group of works for building and general engineering works.
- Construction industry development board Act 38 of 2000
- Prior to this Act, construction industry was not well regulated in South Africa. Therefore, the CIDB Act recognizes that construction industry has an important role in South Africa's economy and expanding infrastructure capacity as a way to advance development. To facilitate sustainable infrastructure development, the Act established CIDB to execute strategy for the reconstruction, development of the construction industry and various matters affecting the industry. According to the Act, the objectives of the CIDB are to:
- Promote the involvement of the construction industry relating to national construction need through the promotion of national, social and economic objectives, industry performance, viability and enhanced value of clients;
- Promote strategic leadership to construction industry stakeholders to encourage sustainable growth, restructuring and development of the construction sector;
- Promote best practice by developing and implementing suitable programmes aimed at best practice and enhanced performance of all parties involved in the construction industry;
- Promote consistent use of policy in the construction industry by all spheres of government;
- Promote growth of the construction industry and the involvement of the emerging sector.

As a way to classify contractors according to the category of works and proper structuring of contractor development in the construction industry, CIDB categorizes contractors according to various levels of grades. In the grading framework, contractors are scheduled in the CIDB in different grades signifying the financial and technical capabilities to execute projects (CIDB, 2009). Table 1 provides the details for tender value range, financial capability requirements, annual turnover, largest contract and available capital for each designated grade. Contractors qualified for lower grades are required to have low turnover and available capital contrary to higher grades contractors.

Table 1: Determining Financial Capability

Designation	Upper limit of tender value range	Best Annual turnover	Largest contract	Available capital
2	R 650 000		R 150 000	-
3	R 2 000 000	R 1000 000	R 500 000	R 100 000
4	R 4 000 000	R 2 000 000	R 1 000 000	R 200 000
5	R 6 500 000	R 3 250 000	R 1 600 000	R 650 000
6	R 13 000 000	R 7 800 000	R 3 250 000	R 1 300 000
7	R 40 000 000	R 24 000 000	R 10 000 000	R 4 000 000
8	R 130 000 000	R 90 000 000	R 32 500 000	R 13 000 000
9	No Limit	R 270 000 000	R 100 000 000	R 40 000 000

Source: CIDB (2009).

Construction industry sector charter

The construction charter on Black Economic Empowerment (BEE) as stipulated in the Government Gazette No. 29616 recognizes the relative isolation of black South Africans from ownership, control and management of productive assets. In the same vein, Lazarus (2007) points out that in order to address BEE in the construction industry, it is necessary to recognize constraints facing emerging contractors. To address various challenges, government introduces CISC to address inequalities in the sector in order to achieve the following:

- To realize transformation in the racial and gender composition of ownership, control and management in the sector;
- To promote the improvement of employment equity in the sector and sticking to appropriate principles;
- To deal with skills development in order to speed-up improved engagement of black people and designated groups;
- To enhance the procurement of goods and services offered by Broad-Based Black Economic Empowerment (BBBEE) enterprises and regulate preferential procurement practices;
- To promote entrepreneurship activities and sustainable growth and development of BBBEE enterprises.

The Act also provides for the achievement of milestones over a specified period of time. Elements of these milestones include ownership, control, employment equity, skills development, procurement, enterprise development and corporate social development by December 2013. Table 2 presents details of the construction charter scorecard with emphasis on weighting and targets in the construction industry.

Table 2: Construction Charter Scorecard

Element	Weighting	Targets
Ownership	25	30%
Management control	10	40%
Employment Equity	10	40%
Skills development	15	1.5% of payroll
Preferential procurement	20	70%
Enterprise development	15	5% of annual turnover
Socio-economic development	5	0.25% of annual payroll
Total BBBEE points	100	

Source: Department of Trade and Industry (2007).

Some of challenges facing small contractors in Polokwane in the Limpopo Province

Although it was stressed out that there have been introduction of various policies by government to promote and support Small, Micro, Medium Enterprises (SMMEs), however, there are still challenges that persist to slow down the development of these businesses to become sustainable businesses. Results obtained from research conducted in Polokwane among SMMEs in the construction industry in 2007 are worth pursuing. In this study, most of the emerging contractors who participated were sustaining their businesses through subcontracting works because of not formally registered with CIDB. Despite this reality, significant constraints were identified (Tshivhase, 2007).

Financial status of SMMEs in the construction industry

Table 3 summarizes the financial status of SMMEs in the construction industry. It shows that 75.76% of SMMEs regarded their financial status as poor to very poor. What can be deduced from this exceptionally high percentage is that these enterprises are generally not financially viable. Furthermore, 24.2% of SM-

MEs indicated that their financial status as good to very good. It is believed that these were basically emerging contractors having CIDB of grade 4 or higher. Further, it is evident that there is indeed a challenge to the management of these enterprises to sustain their business operations. However, failure to register with CIDB may negatively impact on these enterprises since the majority of contractors are not assessed in terms of their financial capabilities and track record. Simplified, their operations may not satisfy financial stress tests required for formal CIDB grading.

Table 3: Financial Status of SMMEs in the construction industry in Polokwane

Status	Percentage
Very poor	21.21
Poor	54.55
Good	21.21
Very good	3.03

Source: Tshivhase (2007).

Management constraints among SMMEs in the construction industry

Table 4 provides an assessment of management constraints facing SMMEs. It shows that 39.1% of participants had deficiencies in financial competencies. Therefore, it clearly demonstrates that SMMEs in the industry have difficulties managing financial matters effectively. Despite this, constraints related to marketing business opportunities accounted to 19.5% which raises a concern with regard to the availability of adequate marketing skills to take businesses beyond their current limitations.

Table 4: Management constraints facing SMMEs in the construction industry in Polokwane

Constraints	Percentage
Financial related issues	39.13
Planning	4.35
Resources	10.87
Marketing	19.57
Technical skills	13.04
Other	13.04

Source: Tshivhase (2007).

Challenges to SMMEs in the construction industry

Table 5 details the challenges facing the SMMEs in the construction industry. It shows that 31.1% of challenges to SMMEs emerged from competition within the industry. Due to lack of key competencies among SMMEs, they are unable to compete effectively. Hoskisson, Hitt and Ireland (2004) add that intense rivalries are normally found in industries with many companies. Nonetheless, challenges associated with lack of projects attained 28.8% illustrating unavailability of opportunities to cater for SMMEs in the construction industry. (Berry *et al.*,2002) have reported that supply-side measures alone are not sufficient for progressive SMME growth if there is a lack of access to potential markets.

Table 5: Threats of SMMEs in the construction industry in Polokwane

Threats	Percentage (%)
Lack of projects or infrastructure investment	28.89%
Competition	31.11%
Nepotism	13.33%
Other (i.e late payments, shortage of skills)	26.67%

Source: Tshivhase (2007).

3. Research Question, Objectives and Propositions

The main research question the study raises is:

- What are the key factors that adversely affect the development of emerging contractors in the construction industry in Limpopo Province?

The study sought to achieve the following objectives:

- To identify key barriers that affect the growth and development of emerging contractors in the construction industry in Limpopo Province.
- To propose feasible and suitable remedial actions that are helpful for supporting emerging contractors in the construction industry in Limpopo Province.

Propositions

- Tshivhase (2007) found that shortage of skills and access to finance remain key challenges to the development of small contractors. The research is thus designed to test what additional factors are more important to construction works in determining the successes of emerging contractors in Limpopo Province.
- Emerging contractors are to a greater extent adversely affected by external environmental or economic factors as opposed to internal factors in the construction industry in Limpopo Province.

4. Study Design, Sample Size and Sampling Techniques

The use of structured questionnaire (Quantitative research)

Structured questionnaires were used to gather data from emerging contractors in the construction industry in Limpopo Province. The design of the study was cross-sectional and descriptive. A random sample of (n=104) emerging contractors was drawn from the 5 districts of Limpopo Province based on stratification.

Methods of data analysis (Quantitative)

The quantitative data was analyzed based on binary (dichotomous) logistic regression analysis (Hosmer and Lemeshow, 2000). In binary logistic regression, the values of each variable of study must be recorded into 1s and 0s. The value of 1 in the study represents existence of barriers whereas the value of 0 represents absent of barriers to growth and development of emerging contractors in the construction industry.

Pearson chi-square test of associations

The Pearson chi-square of association is used for identifying two-way associations among categorical variables. At the 5% level, significant associations have P-values that are smaller than 5% and expected cell frequencies of 5 or above.

Odds ratios obtained from logistic regression analysis

The odds ratio (OR) is termed as the ratio of the odds for x=1 to the odds for x=0 (Hosmer and Lemeshow, 2000). The odds ratio is a measure of association as it approximates likely or unlikely the outcome is present between those with X=1 than those with X=0 whereas Y (dependent outcome) denotes present and absent of barriers to the development of emerging contractors in the construction industry. At the 5% level, significant odd ratios are characterized by odds ratios that are significantly different from 1, P-values that are smaller than 0.05, and 95% confidence interval of odds ratios that do not contain 1.

Hosmer-Lemeshow goodness-of fit statistic

The Hosmer-Lemeshow goodness-of-fit statistic ($\hat{C}$) is calculated by determining the Pearson chi-square from the g x 2 tables of observed and estimated frequencies (Hosmer and Lemeshow, 2000: Ch.5). A formula determining of $\hat{C}$ is given as:

$$\hat{C} = \sum_{k=1}^{g} \frac{(o_k - n'_k \bar{\pi}_k)^2}{n'_k \bar{\pi}_k (1 - \bar{\pi}_k)}$$

where n'_k represents total number of subjects in the kth group, c_k represents the number of covariate patterns in the kth decile,

$$o_k = \sum_{j=1}^{c_k} y_i$$

Where o_k represents the number of responses between the c_k covariate patterns,

$$\bar{\pi}_k = \sum_{j=1}^{c_k} \frac{m_j \hat{\pi}_j}{n'_k}$$

where represents $\bar{\pi}_k$ is the average estimated probability, and the distribution of the statistic $\hat{C}$ is fitted by the chi-square distribution with g - 2 degrees of freedom, x2 (g-2).

The use of in-depth interviews (Qualitative research)

- In-depth interviews were used for data gathering when engaging key stakeholders in the construction industry.
- Tape recorders were used for interviewing 8 respondents.
- Responses obtained were transcribed.
- Tallying, coding and text analysis were used for data analysis.
- Triangulation was used for verifying results of analysis by using subject-matter specialists in the construction industry.
- ATLAS.ti software was used for data capturing and analysis.
-

5. Results of Data Analysis (Quantitative)

Results from cross-tab analysis (Pearson chi-square)
Table 6 shows the results obtained from Pearson's chi-square tests of associations. At the 5% level, significant associations have P-value smaller than 0.05. Table 8 shows that there were significant relationships between barriers and all variables presented. Based on this test, important variables have larger chi-square value. These included variables such as lack of transparency in tender processes (P=0.000), lack of strategy to grow the business (P=0.000), intense competition (P=0.000) and demographic challenges (P=0.000).

Table 6: Pearson Chi-square Tests of Associations (n=104)

Variable of study associated with barriers to emerging contractors	Pearson Chi-square	P-values
Lack of transparency in tender processes	17.7938	0.000
Lack of vision for business	12.949	0.000
Lack of strategy to grow the business	15.913	0.000
Intense competition in the industry	17.530	0.000
Demographic challenges	21.2799	0.000
Payment delays	10.7172	0.001
Lack of Contractor Development Programmes	6.7864	0.009
Poor implementation of BEE policies	5.0816	0.024

Variable "Lack of transparency in tender processes" is further outlined. The results showed that lack of transparency in tender processes has high probability of acting as a barrier to emerging contractors in Limpopo Province. This finding seems to be promoted by doubts about the openness of tender processes leading to the award of contracts. Various attempts may occur in an effort to alter the true outcome of the successful bidder which could act as a serious threat to the development of emerging contractors. This could be manipulated by officials who have authority to undermine the tender processes and appoint undeserved bidders in the Limpopo Province.

Variable "Lack of strategy to grow the business" can be further expanded. It can be stated that lack of entrepreneurial spirit among emerging contractors could lead to poor strategies that are not conducive in growing the business. This may happened as a result of lack of vision for the business even when business is doing well. It is important however to stress out that low level of business management skills could also contribute significantly in terms of failure to develop sound strategies for the business among emerging contractors in Limpopo.

Variable "Intense competition" also emerged as a concern to emerging contractors in the Limpopo Province. It appears that due to larger number of contractors in the Limpopo Province between grades 2 to 7, there are fewer opportunities created to cater for most of these contractors. (Thwala and Mvuvu 2008) state that intense competition in the emerging sector has led to many failures and resulting in unsustainable market. However, lack of regulations to restrict unnecessary entry into the industry also influenced tight competition and this may be confirmed during site inspections.

Variable "Demographic challenges" was identified as one of important variable of study. It showed that emerging contractors are more localized businesses who normally struggle to source work beyond confined districts of Limpopo Province. This may also be impacted by lack of sufficient resources to market businesses in other regions. However, dominance of local firms in regions where emerging contractors are willing to pursue their businesses operations may also pose a challenge. In other instances, due to political pressure, clients may also put restrictions in order to isolate contractors from far especially at local government level.

Results from binary logistic regression analysis

Table 7 shows odds ratios from binary logistic regression analysis. At the 5% level, significant odds ratios have P-values that are smaller than 5%. Table 7 further shows that 4 predictor variables of the 5 variables in the logistic regression are statistically significant at 5% level. These include variables such as lack of marketing business, adverse political environment, lack of efficiency (productivity) and low turnover (lack of projects). These variables have the potential to influence the likelihood of experiencing barrier to the development of emerging contractors in the Limpopo Province.

Table 7: Odds Ratios Obtained from Binary Logistic Regression Analysis

Variable of study for factors that adversely affect emerging contractors	**Odds Ratio**	**P>IzI**	**(95% Conf. Interval)**
Lack of marketing business	8.58	0.001	(2.48; 29.67)
Adverse political environment	15.41	0.002	(2.72; 87.34)
Lack of efficiency (Productivity)	5.52	0.005	(1.67; 18.17)
Low turnover (lack of projects)	4.67	0.011	(1.40; 15.19)
Slow pace of upgrading	2.99	0.076	(0.89; 10.05)

Table 7 shows that the odds ratio of the variable "Lack of marketing business" is equal to 8.58. This means that a contractor who does not market his or her business well is 8.58 times as likely to encounter a barrier in contrast with an emerging contractor who markets the business well. This explains that high rate of failure may come from lack of commitment to formulate marketing plans for the business in consistence with business long-term vision as well to widened market share in the Limpopo Province.

The odds ratio of the variable "Adverse political environment" is equal to 15.41. This means that a contractor who operates in a complex political environment is 15.41 times as likely to encounter barrier in comparison to contractor who is not confronted by adverse political environment in the industry. (Ketchen, Eisner, Dess and Lumpkin 2009) argue that political processes and legislation should promote the enabling environment with which industry must promote. Despite this, situations may be emerged where political power is exercised to divert work opportunities assigned for competent emerging contractors as to fulfil own interests in the province.

The odds ratio of the variable "Lack of efficiency (productivity)" is equal to 5.52. This shows that contractor who is inefficient due to lack of competencies in the construction industry is 5.52 times as likely to encounter barrier contrary to a contractor who performed well and complete the projects within specified time. Therefore, without key expertises in the industry emerging contractors will find it difficult to perform well and achieve positive profit margins.

The odds ratio of the variable "Low turnover (Lack of projects)" is equal to 4.67. This shows that a contractor who achieves low turnover for a specified period is 4.67 times as likely to encounter barrier in comparison to contractor who attained high turnover. This also aligns to the ability of emerging contractor to compete and market projects effectively. Further, high rate of contractors competing for the same work seems to have intensified competition in the Limpopo Province which makes it difficult for one to get a tender.

The odds ratio of the variable "Slow pace of upgrading" is equal to 2.99. This means that contractors who takes more time than expected to upgrade CIDB grade is 2.99 times as likely to encounter barrier in comparison with another contractor who upgrade on constantly. (Cameron 2007) argues that it is risky to think that small companies can upgrades CIDB grades rapidly.

Assessment of goodness-of-fit

The reliability of the estimated logistic regression model is evaluated by classification table and the Hosmer-Lemeshow goodness-of-fit test.

Classification table

Table 8 shows the assessment of model fitness based on classification. Sensitivity is the ability of fitted logistic regression model to accurately predict emerging contractors who are likely of going out of business. Specificity is the ability of the fitted regression model to accurately predict emerging contractors who are doing well. The overall rate of correct classification is equal to 81.73%. Both sensitivity and specificity are above 50% which showed that the fitted model is fairly well sensitive and specific.

Table 8: The Classification Table

Sensitivity	Pr (+ I D)	80.43%
Specificity	Pr (- I~D)	82.76%
Positive predictive value	Pr (+ I D)	78.72%
Negative predictive value	Pr (- I~D)	84.21%
Correctly classified		81.73%

The Hosmer-Lemeshow goodness-of-fit-test

Table 9 presents results of Hosmer-Lemeshow goodness-of-fit-test. To achieve model fitness based on Hosmer-Lemeshow test, P-value should be greater than 5%. In this test, the P-value of 0.7296 was achieved, which demonstrated the reliability of fitted model. The fact that the P-value is greater than 5% explained the soundness of fitted model.

Table 9: Results Obtained from the Hosmer-Lemeshow Goodness-of-Fit Test

Number of observations	104
Number of covariance patterns	25
Pearson chi2 (19)	14.89
Prob > chi2	0.7296

6. Results of Data Analysis (Qualitative)

Lack of sustaining emerging contractors in the Province

According to emerging contractors who participated in this study, there is limited chance of sustaining a business in the industry under current conditions because contractors are unable to get new contracts continuously. The reality of uncertainties regarding future prospects for securing new contracts is also compounded by government financial constraints and subsequent insufficient investments in infrastructure projects. This underscores the requirement to survive for emerging contractors, which is through exploring opportunities where other competitors lacked a particular skill, particularly in specialized works which few contractors undertake. The extent to which emerging contractors may succeed depends on a variety of factors – which they can neither control nor influence.

Lack of subcontracting work to local emerging contractors

Subcontracting is a work practice that seeks to outsource contracts of certain types of work to other companies. It is a practice which is recognized throughout the world (Ohnuma *et al.*, 2000). It is an option often exercised when the primary contractor does not have time or relevant skills to perform required tasks – or as a displacement of mundane tasks that established enterprises may consider mundane.

In Limpopo Province, large construction projects were constructed in the past including preparations for the 2010 FIFA

Soccer World Cup. However, most of these projects did not benefit local emerging contractors. The nature of completion for the allocation of work packages was such that small and emerging contractors were disadvantaged by default. The only prospects for any meaningful benefit from the projects were through subcontracting. According to the (DTI 2003), this remains a critical challenge for small businesses as they consistently have to confront with market conditions that exclude their participation in favour of established enterprises.

Some of emerging contractors interviewed approached established construction companies that were awarded contracts within the province to solicit subcontracting works. Despite this, when opportunities for subcontractors came, work allocations were bestowed on those subcontractors who were not local despite having necessary skills to execute the work packages.

Lack Investment in Infrastructure in Limpopo Province

According to (Milford *et al.*, 2002), the prominent challenge facing the construction industry is increase investment on infrastructure which relies on the economic growth of a country. The persisting international financial crisis continues to limit government expenditure patters and fiscal balance. As such, most governments have either suspended or limited the allocation of civil work packages to prioritize frugal fiscal rectitude. In the Limpopo Province, this has resulted in profound of limited availability of infrastructure projects in the post-2010 FIFA Soccer World Cup period.

In addition to embedded widespread skills shortages, some of the emerging contractors with qualified and experienced personnel are underutilized. There are inherent challenges to this development and as (Berry *et al.* 2002) assert, there is a great risk of investing more resources on a supply side while demand constraints are significant. These limitations arise from the contraction of the global economy due to the financial crisis since 2008 – from which the global economy has not recovered fully. With this reality, a significant number of emerging contractors are marginalized by a rather competitive industry with limited opportunities.

Lack of access to finance by emerging contractors in Limpopo Province

Emerging contractors have increasing challenges in accessing financial resources to undertake their projects (Gauteng 2009). Due to their underdeveloped business operations and often precarious financial standing, emerging contractors experience challenges when seeking financial resources. The tightened requirements for accessing credit have restricted accesses to funding for many emerging contractors. Failure to secure financial support makes it difficult for emerging contractors to run a project, acquire plants and equipments. Compounding the problem further are recurring cash flow challenges, which negatively impact on the consistency necessary for sustained credit accessibility. There is also evidence of unsustainable business practices – to which most emerging contractors appear more susceptible.

Shortage of skilled personnel accessible by emerging contractors in the construction industry

Due to the national skills deficit, the availability of qualified personnel is a primary challenge in the industry. Emerging contractors have to compete with established enterprises to attract qualified personnel. To address this, some of emerging contractors tried to train newly qualified graduates. However, the problems of natural attrition are such that emerging contractors lose such hard-acquired skills due to competitive remuneration packages offered by established contractors. The exposure to relevant advanced experiences offered by established contractors in their projects make them naturally desirable.

7. Conclusion

Despite internal organizational weaknesses, results from both quantitative and qualitative research showed that emerging contractors are significantly impeded by factors originated from external environment. It has emerged that relevant construction skills and capital resources are not the only key determinants in determining successes among emerging contractors in the industry. Therefore, there are number of factors that play a significant role in shaping the growth and development of emerging contractors in Limpopo Province.

Findings from statistical analysis

The Pearson chi-square tests of associations showed that growth and development among emerging contractors is significantly influenced by lack of transparency in tender processes, lack of vision for business, lack of strategy to grow business, intense competition, demographic challenges, payment delays, lack of contractor development programmes and poor implementation of BEE policies in the construction industry. These factors could play an influential role to the extent in which emerging contractors succeed in the Limpopo Province.

Despite challenges emanating due to skill shortages and access to finance among emerging contractors in Limpopo Province, the results obtained from binary logistic regression analysis showed that growth and development of emerging contractors is significantly affected by lack of marketing business, adverse political environment, lack of efficiency (productivity) and inability to achieve reasonable turnover. These variables have likelihood to influence growth and development of emerging contractors in the Limpopo Province.

Findings from interviews

The study established that large-scale projects are not creating opportunities for emerging contractors. This included projects that were undertaken in preparation for 2010 FIFA Soccer World Cup. While the pre-Soccer World Cup created a remarkable degree of optimism for small contractors to benefit materially from the construction process, such expectations turned out to be misplaced. Further, without the prospects of similar projects being undertaken in the near future, the prospects for emerging contractors are not encouraging. The introduction of BBBE status to measure the contribution of established firms to emerging companies remains to be seen. Adherence of such contributions requires a degree of formal measurement to assess the usefulness of this requirement.

The study found that insufficient investment in the infrastructure projects has negatively impacted on the development of emerging contractors in the construction industry. This seems to be further enhanced by infrastructure backlogs that are experienced by municipalities where most of emerging contractors get work opportunities. However, increasing number of emerging contractors in Limpopo Province could also contribute significantly to this since available projects are not sufficient to provide for majority of contractors in the Province.

The study established that access to finance is still a critical issue as far as the development of emerging contractors is concerned. It was revealed that some of provincial agencies that assist emerging contractors use the same criteria when applying for a loan to commercial banks. This was indicated as a challenge to emerging contractors to access a loan. It seems that due to insufficient funds allocated to provincial agencies, this led to stricter criteria for one to be granted a loan.

It also emerged that emerging contractors struggled to access key personnel in various areas of works. It appeared that due to high risk for emerging contractors to sustain a business considering tight economic conditions, this played a central role for these businesses to acquire qualified personnel and offer lucrative packages. It showed that emerging contractors are failing to compete with established contractors and normally lost key personnel in the process. Also, some of contractors reported to capacitate skills to technicians directly from tertiary institutions but when they acquired those skills they start to look for opportunities among established contractors.

8. Recommendations

Coming from the findings are recommendations for policy makers and key stakeholders in addressing challenges facing the development of emerging contractors in Limpopo Province.

A need to marketing business beyond local market or Limpopo Province

The statistical analysis showed that emerging contractors are significantly affected by lack of commitment to market business.

This was further supported by much lower turnover among most of the participants. Moreover, some of emerging contractors were found to have no projects in progress which sent a signal that such businesses are faced with the risks of going out of business. This appeared to be affected by larger number of emerging contractors that could not be sustained by available projects in Limpopo Province. The study therefore recommended that emerging contractors should tender projects beyond the confined areas of Limpopo Province so as to minimize risk of failures.

A need to restrict political interference in tender processes in Limpopo Province

The statistical analysis established that adverse political environment could contribute negatively to the development of emerging contractors in the construction industry which greatly depends on the level of political intervention particularly for government contracts. This may happen when political connected businesses are given preference over competent emerging contractors in Limpopo Province. It is therefore recommended that there should be limitations in which political meddling in tender processes take place and such intervention should only be done outside tender processes aiming to create an enabling environment for the development of emerging contractors in the construction industry consistent with government objectives in empowering small businesses.

A need to enhance efficiency by capacitating emerging contractors

The statistical analysis showed that lack of efficiency have a serious influence in the manner in which emerging contractors execute the work successfully. This seems to be associated with the low level of skills necessary to undertake the work. In this regard, it is recommended that Contractor Development Programmes aiming to capacity emerging contractors in areas of financial management and planning, contractual management and technical skills be intensified in the Limpopo Province

A need to improve access to finance by emerging contractors

The study revealed that emerging contractors in Limpopo Province struggled to access loan in order to finance their projects. This had contributed negatively on the sustained growth among these entities. It appears that lack of sufficient funds led to few businesses assisted while majority of businesses continue to be denied this opportunity. It is recommended that provincial government should lessen requirements for granting financial backing to emerging contractors and be further injected with more funds to address current financial backlogs and able to assist most number of businesses.

A need to empower emerging contractors through subcontractors from larger contracts in theLimpopo Province

The study established that emerging contractors struggled to access subcontractors from larger contractors who undertook major projects within the province. It is recommended that clients should take the leading role by ensuring that local contractors are promoted through providing such requirements in primary contractual obligations. Such contractual obligations however should not compromise the quality and regulations governing civil constructions.

References

Berry, A.L., Von Blottnitz, M., Cassim, R., Kesper, A., Rajaratnam, B. and Van Seventer, D.E. (2002), 'The economics of SMMEs in South Africa: Trade and industrial policy strategies', Available at: http://www.edgegrowth.com.pdf (last accessed 01 November 2010).

Cameron, L. (2007), 'Too few contractors moving up the grades', Available at: http://www.engineeringnews.co.za/article/too-few-contractors-moving-up-the-grades-2007-11-02 (last accessed: 18/10/2001).

Construction Industry Development Board. (2009), 'Application for contractor registration: grade 2 to 9', Form CRSF006, Version 3.6, Pretoria, South Africa.

Construction Industry Development Board. (2003), 'Construction industry development board Act number 38 of 2000', Cape Town, South Africa.

Department of Trade and Industry. (2007), 'Broad-based black economic empowerment act (53/2003): Construction Sector Charter on Black Economic Empowerment', Notice No. 111 of 2007, Government Gazette No. 29616.

Department of Trade and Industry. (2003), 'The integrated small business development in South Africa 2004-2014', Pretoria, South Africa.

Diale, A. J. (2009), 'Government support for Small, Micro and Medium Enterprises (SMMEs) as a venture for business and economic development in South Africa', *Repositioning Africa business and development for the 21st century*, Tshwane University of Technology, Pretoria, South Africa.

Gauteng Department of Public Transport, Roads and Works. (2009), 'Emerging contractor development policy', Johannesburg, South Africa.

Hoskisson, R.E, Hitt, M.A. and Ireland, R.D. (2004), 'Competing for advantage', Ohio, Thomson South-Western.

Hosmer, D.W., and Lemeshow, S. (2000), *Applied Logistics*' 2nd edition, New York: John Wiley and Sons.

Ketchen, D.J., Eisner, A.B., Dess, G.G. and Lumpkin, G.T. (2009), *Strategy*, New York: McGraw-Hill.

Lazarus, S.J. (2007), 'The development and assessment of an integrated skills development model for emerging construction contractors', The University of Free State, Bloemfontein, South Africa.

Marx, S., Von Rooyen, D.C., Bosch, J.K. and Reynder, H.J.J. (1998), *Business Management*, 2nd edition, Van Schaik Publishers, Pretoria.

Milford, R.V.J., Hodgson, S.J., Chege, L. and Courtney, R. (2002), 'Construction industry development: The need for infrastructure investment', Available at: http://www.odsf.co.za/cdcproc/docs/3rd/milford_rv_hodgson_s.pdf (last accessed 13/09/2010).

National Advisory Council on Innovation. (2003), 'The potential impact of skills shortages on the innovative capacity of major capital engineering projects', Department of Science and Technology, Pretoria, South Africa.

Ohnuma, D.K., Periera, S.R and Cardoso, F.F. (2000), 'The role of subcontractors in the competiveness of building companies and the integration of value chains: Subcontracting in construction companies', *Proceedings of the CIB W92 procurement system symposium*, Santiago, Chile, April.

Watermeyer, R.B., Gounden, SM., Letchmiah, D.R. and Shezi, S. (1998), 'Targeted procurement: A means by which socio-economic objectives can be realized through engineering and construction works contracts', *in Journal of the South Africa Institution of Civil Engineering*, 40(4).

Thwala, W.D. and Mvubu, M. (2008), 'Current challenges and problems facing small and medium size contractors in Swaziland', *inAfrican Journal of Business Management*, 2(5): 93-98.

Tshivhase, K.L. (2007), 'An investigation into the management issues that affect the growth and development of Small, Micro and Medium Enterprises (SMMEs) in the construction industry in Polokwane: Limpopo Province', Management College of Southern Africa, Durban, South Africa.

Chapter 10

Forging Partnerships for Innovation through Networking: Case Studies from the Fodder Innovation Project, Nigeria

Elias Madzudzo

1. Introduction

There is consensus among stakeholders that a revitalized agrarian economy is a key turning point in economic and social development of Africa. In other words, innovation i.e. continuous learning for improvements in product design and organizational culture (Mytelka and Smith, 2002; Andrew and Evans, 2011), is a prerequisite for and a pathway to dynamic agricultural systems (Clark, 2002; ICSU, 2002). However, there is also consensus that a disconnect exists between research or supply side products and the user or demand side actors' expectations whereby a weakened innovation capacity emerges (Clark, 2002; Hall and Dijkman, 2009). This weakened innovation capacity is a result of atomistic institutions or ways of working among components of agricultural innovation systems. This study uses case studies from the Fodder Innovation Project (FIP II) to show that when actors come together in dialogue an emergent property is networking. It then goes on to argue that this networking creates opportunities for optimal partnerships between research and development -a process that leverages innovation capacity.Admittedly, there are successes emerging from the use of research, however, especially for developing countries, for each success story there are many missed opportunities and outright failures (ICSU,2002; CGIAR, 2011; Giller *et al,* 2011) or as Rukuni *et al* (1998:1073) put it, '...Africa's farmer support systems are in serious disarray'. The reasons for these failures are several,but two stand out. Firstly, much knowledge remains untapped due to the failure of the research systems to encourage an awareness of systemic linkages and a regard for experimental learning in those whose experiences and behaviours will be central to any transition towards sustainable development (ICSU, 2002; Klerkx and Leeuwis, 2008; Manyong, 2005). Secondly, there is an inverse relationship between potentially valuable science and technology painstakingly produced and its use (Biggs and Clay, 1981; ICSU, 2002; Hall, 2009). The re-alignment of research and development priorities, which is the subject of this paper, is topical globally and locally (CGIAR, 2011; Hall *et al,* 2003; IITA, 2005). There are calls for 'a new contract between science and society' (ICSU, 2002) that will enable, especially,African governments to address the conundrum between development research and especially resource poor users (Rukuni, *et al* 1998). Sayer and Campbell (2004) call for a new role for since where it is a joint venture between scientists and resource managers. Hall *et al* (2003)pose a research challenge for evidence that shows how institutional learning leads to stakeholder driven ways of setting technical research priorities. The CGIAR is responding with a shift from a linear approach that hierarchically sequences research to extension to users to an approach that embeds research in development realities of the users (Sayer and Campbell, 2004; CGIAR, 2011). This paper presents action research based evidence from the FIP II that addresses this challenge. In general, this paper examines the possibilities of leveraging synergies between the demand side and supply side actors within the agricultural knowledge infrastructure. Specifically, the paper focuses on how networking is a key process that connects development back to the research system,or as the World Bank (2006) puts it, embedding research into development through improved innovation capacity.

In this study, it is assumed that development research is relational and context sensitive (Ellis, 1999). It maximizes its potential from interfacing with other ideas, practices and developments within certain spaces and places (Hall and Dijkman, 2009; CGIAR, 2011). We explore and illustrate these arguments using the Fodder Innovation Project (FIP II) implemented in Nigeria.

The FIP II, was a sequel to an earlier Fodder Innovation Project (FIP I). Two key lessons had emerged from FIP I experiences. Firstly, that technologies for livestock fodder are available in Nigeria,for instance the National Animal Production Research Institute (NAPRI) at Zaria, has a long history of developing fodder technologies (ILCA/NAPRI, 1984). Secondly,that these technologies are embedded in a policy and institutional context, which has a bearing on their adoption (Hall *et al*, 2008). FIP II approached the fodder challenge from the view that fodder scarcity is symptomatic of weak innovation capacity. It therefore sought to strengthen innovation capacity as a pathway to addressing livestock fodder challenges and opportunities in Nigeria. Put differently, solutions to the livestock fodder problem are a product of the institutional context of Research and Development (R&D) i.e. interaction or networking of *inter alia*, researchers, private and public sectors, extension and farmers. Outcomes from the FIP II project networks are unusual partnerships between research and development actors that make research resonate with local interests. This significant change in actor habits and practices is the subject of this paper. Illustrations from the field lead us to argue that networking makes research and development work better.

2. Organisation of the paper

This chapter is organized as follows: the next section discusses innovation, networking, partnerships, research and development in light of current theory and policy debates. The major argument is that research and development is always embedded in social, political and institutional contexts (Leach *et al*, 1999). A diagnosis of innovation capacity in the Nigerian agricultural system follows from the diagnosis the hypothesis that networking leads to opportunities for unusual partnerships between research and development is developed. Case studies from two FIP II sites are presented to illustrate the emergence of unusual partnerships as a result of strengthening links and networks. A synthesis of emerging lessons in the discussion and conclusion wrap up the argument.

3. Conceptual framework

As African agricultural systems enter into global relationships, they have become linked to external dynamics - e.g. dependence on external sources of technology, energy and external markets - at varying degrees of engagement (Long, 2000). Global linkages have not removed, but in some cases, accentuated economic and social heterogeneity in the rural agricultural systems (Ellis, 1994). For instance, Nigeria's engagement with international capital in the development of cocoa value chains benefits *inter alia* public servants and trade brokers and further impoverishes producers upstream of the value chain (Manyong *et al*, 2005; Adeoti and Olubamiwa, 2009). A general lesson that emerges from these observations is that though a universal phenomenon, global relations do not have uniform impact everywhere. From this, we draw a cognate lesson that narratives that perceive innovation in terms of 'prime movers' e.g. science and technology, the market or peasants may not be useful in research and development planning in Africa (Allison, 2011; Hall *et al* 2003; Klerkx, 2008). Instead, we assume that there is an organisational culture e.g. atomistic behaviour, inward looking leadership, which can be changed through networks leading to productive partnerships for innovation in R&D.

Innovation, as opposed to invention, encompasses improvements of a technical, institutional and political nature (Hall *et al*, 2003; Mytelka, 2000; OECD, 2000). Innovation is an interactive and iterative process emergent from diverse sources of knowledge. Likewise, agricultural innovation implies embedding technical knowledge in particular normative and policy contexts. The extent to which research is embedded in a context will determine its resonance with user interests and characteristics i.e. innovation capacity. Considering innovation capacity not only widens our perception of research success and failure; but repositions research foci to respond to the complex social and political environments of Sub Saharan Africa (Long and Arce, 2000; Clark, 2002;). For instance, IITA's assessment of constraints in Nigeria's agriculture identifies poor government support as preventing the emergence of innovations from research institutes. Even when the technology is available there are communication gaps between the researchers and the end users (Manyong *et al*, 2005). The manner in which research is embedded in the institutional and political environment and how it impinges on innovation capacity is a key feature of an innovation systems paradigm that we address in this chapter.

The innovation systems paradigm rejects the separation of knowledge producers e.g. scientists, and consumers, e.g. farmers (Clark, 2002; Hall *et ul* 2003, Klerkx and Leeuwis, 2008). Innovation implies complexity, multiple actors including research and development, and co- evolution of scientific, technological and social systems (Sayer and Campbell, 2004; Hall *et al*, 2008); an indictment of narratives that view rural innovation in linear terms, where knowledge producers and consumers are in some binary and sequenced relationship. These technology models are criticized for a linear vision or pipeline approach (Clark, 2002) and for ignoring barriers in the knowledge to action chain. A result of such barriers is that in most developing country economic systems, the scarce resource is the capacity to identify what is relevant and organize it in a productive way; a process that requires co-evolution of knowledge production and use processes (Clark 2002; Sayer and Campbell, 2004).This interaction or networking enhances the innovation capacity of the system to identify, adapt and exploit new or modify old knowledge. Innovation capacity as an emergent property of the interactive process points to the importance of networking in fostering partnerships. Networking fosters a complex system of cognate actors involved in the production, diffusion and adoption of knowledge (Sumberg, 2005; Hall, 2006).

In this study's conception of innovation capacity, networking produces unusual partnerships. Opportunities and challenges in African agriculture increasingly require actors to network with others to optimise on comparative advantages, there is need to go beyond the usual partnerships to optimise on capacities (CGIAR, 2011; Allison, *et al* 2011). The nature and pace of change is such that social and economic arrangements must adjust to changing agendas, priorities, realities, and opportunities. On the contrary, as we go on to illustrate in Table 1 below, in evolving innovation systems as are found in Africa, habits and practices that promote system disconnectedness abound (Manyong, 2005). These 'complementary shortfalls,' as Clark puts it, negate the 'the necessary linkages to ensure healthy technological development' (2002:366). This fostering of unusual partnerships in systems comprising disconnected platforms, we hypothesise In this chapter, is a product of networking that emerges in the search for innovation capacity. Implied in our argument is that these partnerships are not solely or always associated with frontier research but also with incremental, mundane 'under the radar' problem solving that farmers and research organisations undertake to address challenges.

4. Nigerian Agricultural Innovation System Competency

This section presents a diagnosis of the Nigerian innovation system conducted at the beginning of FIP II,in 2008. Since local innovation systems are embedded in wider innovation contexts it suffices here to present generalized assessment of the configuration of the innovation system in rural Nigeria. More detailed diagnoses are presented as part of the cases studies below. The Nigerian rural landscape, like most rural areas in Africa, is characterized by several institutions and organisations.The litany includes different government ministries, various units of the NARS, including small enterprise development, agriculture and extension, local government authorities, Non Governmental Organisations (NGOs), finance institutions, and agricultural input suppliers.

While the existence of these innovation ecologies is evident, *can we assume the same for the partnership making connections?* The historical overview below shows that partnership making connections between components that ensure flow of information and facilitate learning are weak (Adeoti and Olubamiwa, 2009). This disconnect promotes incentives that are averse to the emergence of partnerships that promote innovation capacity.

Current habits and practices found in the Nigerian agricultural landscape are best understood by reference to roles that rural areas have historically played in the economy. Rural areas, in both colonial and postcolonial erashave been sources of particular raw materials. For instance, northern Nigeria has been known for the production of groundnuts as an export crop (Shenton, 1986). The South West was a major producer of cocoa and palm oil for export (Berry, 1975; Walker, 2000). To the national and international economy, these rural areas were best known as suppliers of these export crops or raw materials. An infrastructural and administrative system was established to facilitate the research, production, marketing and transportation of these crops. For example, the only rail system established during colonialism followed the groundnut and cocoa routes. Similarly, a road network system was established to facilitate the extraction of the raw materials from the rural areas (Olubomehin, 2006).

A focus on specific crops in these rural areas led to the development of mono-cropping regimes in response to the state controlled market. Farmers put emphasis on these crops in response to the state that also intensified its investment into specialized research and extension in these sectors (Idachaba, 1980). For example, the Cocoa Research Institute of Nigeria (CRIN) was established specifically to explore means of ensuring that the country maximized on its comparative advantage as raw material provider in the value chain with no commitment to building innovation capacity. The same was the case with other export crops (Adeoti and Olubamiwa, 2009). As illustrated in Table 1 below,sector based clusters with unidirectional relationships were established that encouraged specific research and training organisations to focus on their traditional atomized roles.

Table 1: Innovation Systems in Perspective: Institutional Behaviour across Sectors

Attribute	Export agriculture sector	Subsistence agriculture sector
Product	Sector specific	Mixed
Research and Extension	Formal and commodity specific	Weak later donor supported
Markets	Formal urban and international	Local and informal
Linkages/ Connectedness	Within sector	None/*ad hoc* incomplete
Gender	Masculine	Feminine
Relationship with the state	Strong	Weak informal paternalistic
Learning	Sector loop	Technology transfer
Finance	Producer support formal	Informal exploitative
Ease of entry	Conditional	Open
Innovation Trajectory	Orchestrated	Opportunity driven
Diversity of approach	Mono-cropping	Opportunistic
Decision making	Corporate	Private/ individualistic
Risk	Diffused across sector	Household/ individual level

Source: Compiled by the author

Two broad sectors characterized the rural innovation landscape; an export agriculture sector and a subsistence agriculture sector (Bates, 1981). Each developed habits and practices that impinge on present day innovation dynamics. These habits and practices of partnerships, coordination and mutual learning among actors in Nigeria are shown in Table 1.

Table 1 presents a generalised background to institutional practice in the Nigerian sector. Colonial and post colonial strategies focussed on cash cropping with little regard to enhancing the innovation capacity in the subsistence sector. As a result, the institutions around the production, marketing of commodities from this sector have largely been based on individual farmer initiatives. There is minimal or no interaction with other actors to facilitate collective reading and response to opportunity sets. Agricultural research platforms, thus configured, promoted and made disconnectedness normal. As Manyong*et al.* (2005) observe, the rural livelihoods pathways reflected contours of the actor configurations presented above. The rural landscape at present reflects this original bias towards commodity exports.

In general, Nigeria paints a bleak picture of the rural innovation system in the policy and institutional domains. The post independence rural development policies have continued to parody the colonial era where rural areas were largely sources of specific raw materials supported by a commodity extension and research model e.g. cocoa (see Mustapha and Meagher, 2000; Adeoti and Olubamiwa, 2009). Institutionally, the actors that comprise the innovation systems in Nigeria hardly interact; informal networks exist but simply translate into disconnected platforms where learning and collaboration barely take place, in other words, weakened innovation capacity. Table 2, below is a summary of these habits and practices that act as barriers to innovation capacity in the agricultural sector of Nigeria.

Table 2: Constraints to Effectiveness of Agricultural Policy in Nigeria

Policy Instability	High turnover in agricultural policies, policies formulated and scrapped in rapid succession.
Inconsistency in Policies	No *systems approach* to policy formulation each problem viewed in isolation; Government agricultural policies and industrial programs mutually antagonistic e.g. conflict between government's domestic food production policy and its cheap food import policy.
Narrow Base of Policy Formulation	Actors at the grassroots alienated from policy formulation; policies lack grassroots support.
Poor Implementation of Policies	Little attention paid to the efficient implementation of policies; red tape, corruption.
Weak Institutional Framework for Policy Coordination	Inadequate institutional arrangements for policy and program coordination often leading to duplication of effort in resource use among units of the same government.

Source: Compiled from Manyong et al., (2005).

Table 2 shows the barriers to innovation capacity in the agricultural sector of Nigeria. As such, it may be faulted for the static picture it depicts of what is essentially a dynamic system. Indeed, generally the World Bank (2006) and specifically for Nigeria, Manyong *et al* (2005) and Adeoti and Olubamiwa (2009) among other Nigerian policy analysts report efforts by the Nigerian state to improve the agricultural sector, for instance the cocoa rebirth initiative. However, the fact that such remedial measures are being undertaken after so many years of research led technology transfer brings the following lessons into relief: firstly, technology development is but a component of a complex process of technology production, transmission and use (Clark, 2002); secondly, the institutional and policy contexts are critical in determining whether the technologies are transformed into innovations that have welfare consequences and thirdly, standardizing technological development is counter intuitive given the context sensitivity of innovation (see also Allison, 2011).

This far, the discussion has raised several analytical dimensions on networking based partnerships that will be illustrated empirically in the cases studies that follow. First, we provide a brief background to the FIP II.

5. The Fodder Innovation Project (FIP II) Context

The FIP II comprises two experiments that examine the set of actor relationships or architectures required to address particular innovation capacity challenges in the livestock economy in Nigeria. While similar in institutional focus, each experiment addresses specific fodder challenges for particular social and ecological contexts (Reddy *et al*, 2010). The steps leading to identifying these experiments included the design of a conceptual framework (Hall *et al*, 2007), which was followed by a landscaping exercise to help select key partners organisations (KPOs) for the project in Nigeria. Institutional histories of the partners, and rapid diagnoses in the target areas provided the bases for the themes that led to the identification of actionable challenges for experimentation. Among the KPOs identified was Sasakawa Global 2000 (SG2000) and Justice Development and Peace Commission (JDPC) respectively, an international and a local non-governmental organisation (NGO) in Nigeria.

A variety of coaching encounters between the research team and the KPOs was designed to help critically interpret, share the idea of innovation system and notion of innovation capacity as resulting from networking linkages and ways of working that make these arrangements to function. These encounters also led to adaptive refocusing of the FIP II action plans based on experiences in the field. KPOs were not given blue prints of how to set up innovation networks, instead project staff coached partners in the development of work plans to help introduce and operationalize these perspectives on innovation and capacity. After that, it was largely to KPOs to implement plans as they best saw. The project continued to interact with KPOs providing advice where it was required over a period of about two years. However, activities continued to take a momentum of their own, shaped by the nature of the KPOs and their larger scope of work, in which they engaged. The experiences from KPOs activities are the bases of the case studies that follow.

Case studies of unusual partnerships between research and development

The following two case studies show that networking within an innovation system, leads to institutional change, which begets innovation capacity. Here institution refers to the practices of actors dealing with research and development (Hall *et al*,2003). Changes in behaviour are illustrated by partnerships and learning between actors towards execution of tasks. The case studies presented here are a continuum of partnership contexts. On one end, the JDPC case, we focus on an actor involved in the production and transfer of animal health technologies, on the other end, the SG2000's case, actors involved in the transfer and use of knowledge. We show that organisational cultures underwent a transformation because partnerships that resonated with a particular institutional context and flexible enough to respond to emerging challenges and opportunities developed. We also use the evidence to further argue that once networks are in place the impact is not restricted to particular spatial and temporal project contexts, but spreads beyond the point of intervention (see also Hall *et al*, 2007). First, we look at the JDPC initiatives.

The experiment in southwest Nigeria focused on intensification of goat production as a pathway out of poverty for those households whose livelihoods are linked to the livestock value chain. The initial situation analysis of livestock, fodder and fodder innovation development showed that households rear small ruminants at a subsistence level. This specialisation and scale is partly due to agro-ecological conditions of this humid zone and more importantly the institutional history of the agricultural sector in the country stated in the background section above. As a result, in setting out technical and institutional objectives, JDPC chose to focus on "transition from subsistence level goat farming to semi-commercial market-driven enterprise". Especially on the institutional dimension, analyses showed that goat production in the south west is a reflection of how policy and institutional dynamics have shaped the rural innovation capacity landscape. Key institutions are the state, the private sector and civil society operating in an atomistic manner. The action research question guiding the JDPC case was, *"how to strengthen the capacity and coordination of existing and potential networks and mechanisms to support the transition of goat farming from a subsistence level to semi-commercial market driven enterprise."* Below, is a case study of the initiatives undertaken to build a network that would enhance information flow.

Networking leads to client oriented technical support

JDPC invited the Nigerian Veterinary Research Institute (NVRI) to be part of the network addressing transition from subsistence to commercial production of goats in Ikire, the FIP II site. To bring out our point of how this encounter led to unusual partnerships we briefly caricature the NVRI as an actor in this network. The NVRI is part of the NARS with a research and development mandate in all aspects of animal health including vaccine production and extension for livestock farmers. Staff in the NVRI is subject to the intercalary tensions of being a research and at the same time a public service bureaucracy. In civil service politics, the latter role is more dominant than the former; the metrics for performance tilt in favour of one's conduct as a bureaucrat than researcher. Such incentive dynamics promote resource allocation mechanisms that resonate with professional expediency than research *per se*. As a result, this limits the scope for responding to demands for research, we illustrate this further below.

Veterinarians find practice in the rural areas challenging because, among other reasons, quacks have infiltrated the profession (Idowu and Babalobi, 2010). Rural veterinary quackery is an institution embedded in a disconnected system characterised by poor state service delivery in general. Specifically, the system dysfunction is a nexus of weak monitoring procedures, contradictory import policy, porous borders that facilitate movement of contraband drugs, and generalised poverty in the farming population. In this context, quack veterinarians compete against the formal sector veterinarians. Buoyed by the systemic paralysis outlined above, quacks naturally emerge winners. The result is that the veterinary officers shy away from any practice that promotes quackery by limiting access to their intellectual property. For instance, veterinarians remove labels on the containers or repackage the chemicals into different containers to restrict the amount of information quacks could use (see also:Turkson and Brownie, 1999). Therefore, veterinarians frown upon initiatives that seek to capacitate non-professionals on any aspect of animal health; *'if you give them an inch they will take a mile'* complained one veterinary officer.

The context presented above shows that there are habits and practices that negate trust building between the research organisation and the users of the technologies leading to weak innovation capacity in the animal health domain. With limited incentives for research, the veterinarians' response capacity to address research demands from within and from outside the organisation is limited. In other words,such a gap implies that users are denied access to research that they can put into use; this is an impediment to innovation. The corollary of this gap is that, the context between the veterinarians and the quacks costs other actors access to useful information and skills (Idowu and Babalobi, 2010). Anyone outside the professional circles is deemed a potential quack. In short, linkages have deadlocked between the veterinarians and the other actors. The capacity of the actors to engage intechnology development has been severely limited. Below we show that the situation reversed through networking,which created opportunities for unusual partnerships.

JDPC invited the NVRI to be part of the FIP network and to participate in the stakeholder meetings. Local government extension workers, FIP Research team, farmers and marketers were also present. The mere presence of other official actors made it appear normal for officials to interact with farmers. Through JD-PC's steering, emphasis was laid at the prospects of transforming the orientation of goat production among the villagers and that all those present were necessary in effecting this transition. Each of the players presented the skills and resources that they could provide. The NVRI representative also mentioned that they were mandated to ensure animal health and, particularly these days (2010) of increased zoonoses like, H1N1 (swine) flu and avian fluto practice strict animal disease surveillance. The farmers used the platform to raise veterinary challenges they experienced as goat producers, which the NVRI provided answers. In the process, it turned out that other players like the goat marketers had tacit knowledge ongoat health, for instance, the herbs to use in times of illness and breeds that are susceptible to illnesses. Farmers shared their experiences with the use of locally available herbs. Far from being the scheming farmers seeking to extract technical knowledge in order to avoid professional practice, they were willing to work with the 'doctor' as a supplement to the knowledge they already had.

Farmers used the networking opportunity to inquire about their goats suffering from a ruminant flu, which they called 'catarrh'. The NVRI officer indicated that the cited problem was *Pestedes Petits Ruminants* commonly called PPR, which is endemic in the area. A vaccine had been developed by the NVRI but there was a 'vaccine break' in this part of the country. This meant that there was no program to supply the vaccine in the area from the head office where stocks are kept. The head office argued that there was no demand for the technology except from the universities. The universities had their own means of sourcing the vaccine. Farmers argued that first, they were unaware of the existence of the NVRI, and secondly, that they were the custodians of the protection against 'catarrh', and lastly, that they wanted the NVRI to provide the PPR vaccine because it was necessary if they were going into commercial production. Later, the NVRI officer was able to contact his head office and a consignment of the vaccine arrived in the area for public use, the first in two years. Other goat producers not directly related to FIP II also benefited from the availability of the vaccine.

JDPC coordinated animal husbandry upgrading by networking the farmers, the marketers, as well as the extension workers from the state Agricultural Development Programme (ADP). For instance, farmers experimented with goat confinement as opposed to free ranging. For confinement to be successful there was also a need to upgrade disease monitoring and control systems among the herds. Yet, extension delivery was stymiedby the paralysis in government institutions and proliferation of quackery and contraband medicines. A remaining option was to train the farmers in primary animal health care and disease management. JDPC consulted the NVRI and invited them to facilitate in the training of the farmers. Based on previous interactions NVRI used this event as an opportunity to fulfil its mandate *viz*, firstly, to weaken the grip of the quacks from the grass roots communities by empowering the farmers through strengthening their links in the research to action chain and secondly, to control disease prevalence through heightened awareness among farmers. The NVRI became the focal point of the farmer training; a significant change from its apprehension before interacting with the farmers.

Farmers reported that the rates of growth among the goats were not commensurate with the quantities of feed consumed by the animals. However, from observations and interviews with the farmers the NVRI officer felt that the problem lay in compromised nutrient uptake due to parasites in the alimentary canal. He recommended that the farmers embark on a de-worming regime urgently to reduce parasite pressure in the animals. NVRI was present at the de-worming camp organised in the village. An unexpected outcome was that the camp attracted even more farmers than those participating in the FIP II. This was a positive case of free riding for NVRI because diseases control coverage was extending beyond the boundaries of the experimental site. Through networking the NVRI was able to upscale its technologies. Another unusual outcome of networking is discussed in the next paragraph.

This episode illustrates two points: as development progresses it will require research but the policy environment might fall short in providing solutions. The second point is that, networking can provide partnerships between research and development to address such demand derived matters. Another unexpected outcome involves NVRI and JDPC collaborating in research. NVRI found a reliable partner to do research with and JDPC benefited from participating in research and also enhancing disease surveillance on its farm. NVRI coordinates an ILRI research on African Swine Fever (ASF) surveillance project (ILRI n.d.). Through networking opportunities for partnerships that integrate development entrepreneurial and research agendas have emerged and gone beyond goat farming.

Better networks timely vaccination in Kano

In 2009, there was an outbreak of PPR in Rogo local government area. At the same time, the local government had given women loans to start goat raising projects. This meant that the PPR would scuttle the local authority's efforts, especially with the understaffed animal health departments (ILRI, 2008). The local government chairperson summoned all the state animal health specialists to attend to the outbreak. He directed that all the animals be vaccinated and affected animals treated in a week. The SG2000 coordinator proposed to him that to expedite vaccination programme they need to include the para veterinarians in the village as well. Initially the local government head was apprehensive that the para veterinarians would not do the job well since they are not directly under his control. The coordinator suggested that the head meets the para-veterinarians before their engagement. Each of the para veterinarians had to bring certificates to prove his or her qualifications as animal health practitioner. This was also a strategy to make sure that the quack veterinarians do not infiltrate the group In the end the vaccination camp comprised 10 para-veterinarians and seven local government and state government animal health specialists.

The outcome of this partnership was timely vaccination that reduced mortality in the local head. The timeliness in conducting the vaccination camp reduced by about 70% the time local government veterinarians would have taken to vaccinate the entire local government herd. In addition, the partnership saved the vaccine from spoilage in this area where refrigeration is not readily available due to erratic electricity supplies. This partnership has demonstrated the value of networks and partnerships to local government authorities.

We also want to show that this case study is an example of evidence based institutional change. For instance, what started off as a specific activity at project level involving FIP II led to system level upgrading as habits changed on the part of the local government. For, when the local government head saw the value of collaborating with the para - veterinary officers he committed himself to invite them in cases where the local government wanted to attend to an animal health issue urgently. According to the SG2000 coordinator who brokered the partnership, the local government head, though initially apprehensive, resolved to budget for similar networks with the para veterinarians. He was considering using the para veterinarians in the annual contagious bovine pleura pneumonia (CBPP) vaccination.

An additional benefit of the network created in the FIP II is the rapid movement of information on animal health. The para veterinarians used in the *Peste Des Petits Ruminants* (PPR) vaccination were quick to alert the authorities of what appeared to be signs of avian flu. The Ministry of Agriculture sent specialists who diagnosed it as chronic respiratory diseases among poultry.

In summary, as aresult of working together the animal health stakeholders formed an unusual partnership that leveraged their reach and resources; these networks helped them avert a crisis. The successful collaboration helped resource poor small goat owners survive because of the synergy emerging from the collaborative efforts ofthe players. In the next example, we take this aspect of synergy and unusual partnerships further.

Implications for R&D

The case has shown how a new a way of interaction with partners has allowed the local government to consider new ways of operating *vis a vis* the para-veterinarians. The subtle institutional context in which technology is introduced matters. Networking-facilitated novel combinations of resources and expertise leading to better delivery of research and technology. These partnerships are low key to the point of being ignored as too modest, yet mundane as they are, they are the ones that respond to and reach the ordinary farmers that form the majority of the rural poor. Such an observation begs the question how do we then scale up such under the radar outputs for impact? The generic knowledge emerging from such initiatives revolves around the principles of fostering innovation capacity. These are: viewing research outputs as time and place sensitive, considering innovation as an emergent process of networking and that this networking will result in partnerships that are not solely headline grabbing activities, in their small way leveraging livelihoods through more responsive and relevant research. More importantly synchronic analysis is inadequate for understanding the process through which development articulates research or vice versa but longitudinal action research methodologies are required to understand the process.

6. Discussion

This chapter argues that networking facilitates the social capital for unusual partnerships between research and development. These unusual partnerships are valuable for facilitating a response capacity that normally slips 'under the radar' yet a critical platform for nesting research in development. Actors are working together on fodder issues in a manner they would not as individual actors. For instance, when JDPC is stuck at the policy level because there is no policy that facilitates response to development derived programmes, an opportunity emerges from a partnership with the NVRI. These partnerships are desirable because they are a turning point in habits and practices leading to trust and reciprocity.

Without networking, resources and expertise held by each of the actors is put to limited use, because of the inward looking institutions that shape their habits and practices. For instance it has been shown that the effort and resources invested in research are not justified by matched demand or use. This has reinforced disenchantment with the technology transfer paradigm whose linear view on innovation is now challenged. Given such a background, the absence of system making connections is glaring. This chapter used examples from experiments done in Nigeria to show that networking provides opportunities for partnerships that facilitate research topic identification, which in turn, leads to knowledge that is usable.

While Nigeria has well established research actors in the agricultural sector in general they have not been able to invigorate innovation in the sector. For example, the under development of the cocoa value chain, is an indictment of current research and development impacts born out of a disconnected innovation system. For the rest of this section we give a synthesis of emerging lessons as well as challenges that may inform and refine policy.

Firstly, networking allows for actors to be aware of other actors in the field

Networking is potentially remedial; it corrects and improves instances where innovation capacity is lacking. In the first case study, the NVRI realised that it was not the only organisation interacting with the farmers in the stakeholder meetings. It was also an opportunity for the farmers to meet *bona fide* veterinarians; the interaction generated social capital that could be used against quacks. In the second case the local government's awareness of the capabilities of the para-veterinarians enhanced its capacity to deal with emerging animal disease with positive outcomes for the poor. Capability to learn and build new competencies depends on how the parts fit together and the strengths of the connections 'goodness of fit', which networking seems to provide in the cases presented here. However, a caveat to these observations is the fact that these relationships require time and other attendant incentives to compensate for seemingly opportunity costs especially in the early stages.

Second, networking articulates research and development

In the JDPC case, collaboration between actors led to stakeholder dialogue, especially between JDPC and NVRI. This partnership facilitated the provision of the PPR vaccine in the area. Supply had been discontinued in the last two years, because it was assumed there was not enough demand. In another instance, JDPC established a pig farm in line with its community development and training objectives. NVRI, on the other hand, was involved ASF assays that required a litter whose case histories could easily be established. Through the networking in FIP II these actors were able to dialogue and effect a strategic partnership. In the SG2000 case the local authority learnt how to optimally use the technical knowledge of the para-veterinarians. In sum,networking leads to institutional change whichleverages greater developmental impact.

Thirdly, these unusual partnerships are not limited to project evaluation spatial and time frames or donor funding cycles

Networking leads to multiple impact pathways as well as to unexpected outcomes. In the first case, the partnership between the actors was not limited to goat farming, but extended to ASF surveillance, which was not in the original plan. In the second case, partnerships heightened animal disease awareness and reporting.

Fourthly, Networking with policy backing is essential

An aspect of both cases is the scale at which they happened. In both cases it was the initiative and intuition of actors. To upscale these partnerships,they need to be supported by a leadership and enabling policy environment. As an institutional change process, to depend on the predilections of certain actors may lead to sustainability uncertainties.

7. Conclusion

The case studies presented above show that building networks leads to productive partnerships among actors. These partnerships are productive in that they address the challenge especially in developing countries of a mismatch between research identification procedures and development priorities; articulation of research and point of use. Networking provides opportunities for dialogue between research and development actors and to explore possible synergies. Inevitably, the process of network building and partnership will necessarily be long and slow because it is about changing habits and practices of engagement that have been nurtured under different, if not perverse, incentive dynamics. At the same time networking and the ensuing partnerships are context sensitive and cannot be speeded up by regimentation or use of blue prints.

References

Adeoti, J. and Olubamiwa, O. (2009), 'Towards an Innovation system in the traditional sector: the case of the Nigerian cocoa industry.' in *Science and Public Policy*, 36(1): 15-31.

Allison, E.H. (2011), 'Aquaculture, fisheries, poverty and food security', *Working Paper, 2011-65*, The World Fish Center, Penang, Malaysia.

Allison, E.H., Bene, C. and Andrew, N. (2011), 'Poverty Reduction as a Means to Enhance Resilience in Small-scale Fisheries', In: Pomeroy, R.S. and Andrew, N. (eds.), *Small-scale Fisheries Management: Frameworks and approaches for the developing world,*Wallingford: CABI.

Andrew, N.L. and Evans, L. (2011), 'Approaches and frameworks for management and research in small-scale fisheries'. In: Pomeroy, R.S. and Andrew, N. (eds.), *Small-scale Fisheries Management: Frameworks and approaches for the developing world,*Wallingford: CABI.

Arce, A. and Long, N. (2000), 'Reconfiguring modernity and development from an anthropological perspective', In: Arce, A. and Long, N. (eds.), *Anthropology, Development and Modernities*, London: Routledge.

Baker, W. (1990), 'Market Networks and Corporate Behaviour', in *American Journal of Sociology*, 96(3): 589 - 625.

Bates. R.H. (1981), *Markets and states in tropical Africa: The political basis of agricultural policies*, Berkeley: University of California Press.

Berry, S. (1975), *Cocoa, Custom and Socio-Economic Change in Rural Western Nigeria,*Oxford: Clarendorn Press.

Biggs, S.D. and Clay, E.J. (1981), 'Sources of innovations in agricultural technology'. in*World Development,* 9(4): 3121-336.

CGIAR (2011), 'Aquatic Agricultural Systems Program Proposal', Available at: www.worldfishcenter.org/.../CRP1.3_Final_4_March_2011.pdf

Clark, N. (2002), 'Innovation Systems, Institutional Change and the Knowledge Market: Implications for Third World Agricultural Development', in Economics of Innovation and New Technology, 11(4-5): 353-368.

Ellis, F. (1999), 'Rural Livelihood Diversity in Developing Countries: Evidence and Policy Implications', *ODI Natural Resource Perspectives,* 40, London: Overseas Development Institute.

Giller, K.E., Tittonell, P., Rufino, M., Wijk, M. van, Zingore, S., Mapfumo, P., Adjei-Nsiah, S., Herrero, M., Chikowo, R., Corbeels, M., Rowe, E., Baijukya, F., Mwijage, A., Smith, J., Yeboah, E., Burg, J. van de, Sanogo, M., Misiko, M., Ridder, N. de, Karanja, S., Kaizzi, C., K'ungu, J., Mwale, M., Nwaga, D., Pacini, C., and Vanlauwe, B. (2011), 'Communicating complexity: Integrated assessment of trade-offs concerning soil fertility management within African farming systems to support innovation and development,' in *Agricultural Systems,* 104(2):191-203.

Hall, A, Sulaiman R.V., Clark, N and Yoganand, B. (2003), 'From measuring impact to learning institutional lessons: an innovation systems perspective on improving the management of international agricultural research', in *Agricultural Systems,* 78: 213-241.

Hall, A. (2006),'Public Private Sector Partnerships', In: *An Agricultural System of Innovation: Concepts and Challenges,*Fodder Innovation Project Working Paper Series 2006-002.

Hall, A., Sulaiman, R. and Bezkorowajnyj, P. (2008), Reframing Technical Change: Livestock Fodder Scarcity Revisited as Innovation Capacity Scarcity, Maastricht: ILRI/UNU-M.

Hall, A. and Dijkman, J. (2009), 'Will a time of plenty for agricultural research help to feed the world?' *Link News Bulletin* Nov-Dec. Available at: http://www.innovationstudies.org

ICSU (2002), *Science and Technology for Sustainable Development,* International Council for Science Series on Science for Sustainable Development, No. 9.Part 1: Consensus Report of the Mexico City Synthesis Conference, Paris.

Idachaba, F.S. (1980), *Agricultural research policy in Nigeria,* International Food Policy Institute.

Idowu, O. S. and Babalobi, O.O (2010), 'Animal Health Management Perspectives Of Rural Livestock Farmers In Southwest Nigeria: The Place Of Community Based Animal Health Workers', in *Nigerian Veterinary Journal,* 31(1): 26-36

IITA (2005), *Bridging the Communication Gap between Scientists and Farmers in Katsina State of Nigeria,* Ibadan: International Institute of Tropical Agriculture.

ILCA/NAPRI, (1984), *Symposium on Livestock Production in the Sub-humid Zone of Nigeria,* Nairobi: International Livestock Research Institute.

International Livestock Research Institute (2008), 'The animal health surveillance system and avian influenza in Nigeria: early detection, reporting and surveillance', Avian Influenza in Africa (EDRS Project), Nairobi.

Klerkx, L. and Leeuwis, C. (2008), 'Balancing multiple interests: embedding innovation intermediation in the agricultural innovation infrastructure', in *Technovation,*28(6): 364-378.

Leach, M., Mearns, R., and Scoones, I. (1999), 'Environmental entitlements: dynamics and institutions in community-based natural resource management', in *World Development,* 27(2): 225-247.

Long, N. (2000), 'Exploring local/global transformations,' In: Arce, A. and Long, N. (eds.) *Anthropology, Development and Modernities,* London: Routledge.

Manyong, V.M., Ikpi, A., Olayemi, J.K., Yusuf, S.A., Omonona, B.T., Okoruwa, V. and Idachaba, F.S. (2005), *Agriculture in Nigeria: Identifying opportunities for increased commercialisation and investment*, Ibadan: International Institute of Tropical Agriculture.

Mustapha, A.R. and Meagher, K. (2000), 'Agrarian Production, Public Policy and the State in Kano Region, 1900-2000',*Drylands Research Working Paper 35*, Crewkerne, United Kingdom.

Mytelka, L. (2000), 'Local Systems of innovation in a globalised world economy', in *Industry and Innovation* 7(1): 15-32.

Mytelka, L.K. and Smith, K. (2002), 'Policy learning and innovation theory: an interactive and coevolving process', in *Research Policy* 31(8-9): 1467-1479.

OECD, (2000), *Knowledge Management in the Learning Society*, Paris: OECD.

Okoli, I.C., Tamboura, H.H. and Hounzangbe-Adote, M.S. (2010), 'Ethno-veterinary medicine and sustainable livestock management in West Africa', In: Katerere, D. and Luseba, D. (eds.), *Ethno-veterinary Botanical Medicine Herbal Medicines for Animal Health*, CRC Press.

Olubomehin, O.O. (2006), 'The development of roads and road transportation in South-Western Nigeria, 1906-1920', in*Nigerian Journal of Economic History*, 4: 14-25.

Reddy V.T.S, Puskur, R., Hall, A., Sulaiman, R, Madzudzo, E., Beshah,T and Bezkorowajnyj, P. (2010), 'Steps to action in FIP II: operational details', *Paper presented at a conference of the Fodder Innovation Project (FIP) II*, 5-6 May, International Crops Research Institute for the Semi Arid Tropics, Hyderabad.

Rukuni, M, Blackie, M.J. and Eicher, C.K (1998), 'Crafting Smallholder –Driven Agricultural Research Systems in Southern Africa', in*World Development*, 26(6): 1073-1087.

Sayer, J. and Campbell, B.M. (2004), '*The Science of Sustainable Development: Local livelihoods and the global environment*', Cambridge: Cambridge University Press.

Shenton, R.W. (1986), *The Development of Capitalism in Northern Nigeria*, Toronto: Toronto University Press.

Sumberg, J. (2005), 'Systems of Innovation Theory and the Changing Architecture of Agricultural Research in Africa', in Food Policy, 30(1): 21-41.

Turkson, P.K. and Brownie, C.F. (1999), 'Perceived Constraints to Privatization ofDelivery of Veterinary Services in Ghana', in *Tropical Animal Health and Production* 31(2): 103-114.

Walker, E.A. (2000), 'Structural Change, the Oil Boom and the Cocoa economy of South Western Nigeria 1973-1980s', in Journal of Modern African Studies, 38(1): 71-87.

World Bank (2006), *Enhancing Agricultural Innovation: How to go beyond the strengthening of Research systems*, Economic Sector Work Report, Washington D.C.: The World Bank.

Acknowledgement

Field data were gathered by the author as a Post-Doctoral Fellow at the International Livestock Research Institute, West Africa Regional Office, Ibadan, Nigeria.

Chapter 11

Assessment of Technological Innovations in Selected Indigenous Oilfield Servicing Firms in Nigeria

Oluseye O. Jegede, Abiodun A. Egbetokun & Willie O. Siyanbola

1. Introduction

Nigeria is the 10th largest oil producer in the world, the third largest in Africa and the most prolific oil producer in Sub-Saharan Africa. The Nigerian economy is largely dependent on its oil sector which supplies 95% of its foreign exchange earnings. The upstream oil industry is the single most important sector in the economy. According to the 2008 BP Statistical Energy Survey, Nigeria had proved oil reserves of 36.22

billion barrels at the end of 2007 or 2.92 % of the world's reserves. The Nigerian government planned to expand its proven reserves to 40 billion barrels by 2010. Nigeria's wealth of oil makes it most attractive to the major oil-multinationals, most of which are represented in Nigeria, with the major foreign stakeholder being Shell. Nigeria produced an average of 2355.8 thousand barrels of crude oil per day in 2007, 2.92% of the world total and a change of -4.8 % compared to 2006 (BP Statistical Energy Survey, 2008).

According to the BP Statistical Energy Survey (2008), Nigeria had proved natural gas reserves of 5.29 trillion cubic metres, 2.98% of the world total. Due, mainly, to the lack of a gas infrastructure, 75% of associated gas is flared and 12% re-injected. Nigeria has set a target of zero flare by 2010 and is providing incentives for the production and use of gas. The government also planned to raise earnings from natural gas exports to 50 percent of oil revenues also by 2010. It has been reported in the BP Statistical Energy Survey (2008) that Nigeria had natural gas production of 34.97 billion cubic metres, 1.18% of the world total.

Nigeria's downstream oil industry is also a key sector including four refineries with a nameplate capacity of 438,750 bbl/d (BP Statistical Energy Survey, 2008). Problems such as fire, sabotage, poor management, lack of turnaround maintenance and corruption were responsible for. The refineries often operate at 40% of full capacity, if at all. This has resulted in shortages of refined product and the need to increase imports to meet domestic demand. Nigeria has a robust petrochemicals industry based on its substantial refining capacity and natural gas resources. The petrochemical industry is focused around the three centres of Kaduna, Warri and Eleme.

The business environment of the 21st century is proving to be very different from the past. This new competitive landscape has greater uncertainty, dynamism and volatility caused by increasing technological developments and the increasingly rapid diffusion of information and new technologies (Chesbrough, 2003; Hitt *et al.*, 2000; Drucker, 1999). This accounts for why some firms are able to implement and use advanced and appropriate technologies with relative ease while others struggle. The link between building local technological capacity and the ability to respond to challenges are usually brought about in the process of learning and co-operation between institutions. The spread of multinational corporation activity around the globe has benefited industries. The trend towards vertical disintegration of production activity in industrialised countries means that developing countries can now participate in the global economy by inserting themselves into international value chains. This is not easy though, as competition to host multinational oil companies' activity in developing countries is becoming tougher. Indeed, evidence shows that only a handful of indigenous oil firms have contributed to the productivity in the sector. Even-though, there is a level of Foreign Direct Investment (FDI) in Nigeria's oil and gas sector however, indigenous firms have not benefited from the internationalisation of production. According to Ozigbo (2008) the local content in the industry is still very low as over 60 percent of the work value in the oil and gas sector is carried out abroad.

Several studies have assessed the level of interaction among the various elements of Nigeria's National Innovation System (NIS) and analysed the factors influencing the generation and successful adoption or non-adoption of industrial innovation by Nigerian firms (Adjebeng-Asem, 1996; Oyewale, 2003; Oyebisi *et al.*, 1996; Oyelaran-Oyeyinka *et al.*, 1996). There is also a large body of evidence on the innovative characteristics of the manufacturing sector in Nigeria, particularly the small and medium enterprises (SMEs) (e.g. Abereijo *et al.* 2007; Oyelaran-Oyeyinka, 1997, 2001, 2002, 2005; Oyelaran-Oyeyinka *et al*, 2007; Oyelaran-Oyeyinka and Lal, 2006). In literature, there is very little attention given to the intensity of innovation in the indigenous oil and gas servicing firms that are providing services to the foreign oil and gas firms in Nigeria. Furthermore, the system approach was used in virtually all of the studies, which is good especially for the developing country context, but little attention is paid to sectoral influences on firm-level innovation capability. In addition, most studies have narrowly focused on either product or process innovations; not much has been said concerning other types of innovations. Indeed, there are not many detailed sub-sectoral studies that can be identified in the literature on innovation in Nigeria's indigenous oil and gas servicing firms.

In developing countries and regions, the process of industrial development involves the employment of extant technologies to overcome the wide gaps between them and the industrial forerunners (Oyelaran-Oyeyinka, 1998). Thus, to close the wide technological gap between the developed and the developing countries, the later need to adequately develop, deploy and update the existing technologies, as well as strengthen necessary technological capabilities and knowledge required for technological and innovative development. As noted by Oyelaran-Oyeyinka (2004), another strategy for reducing technological disparity between the developed and developing countries is how new technology is acquired and adapted. This is, through technological acquisition. Technology is a term used for not only physical devices but also encompasses ideas, procedures, know-how and social institutions involved in the technological processes; all of which are inseparable elements/factors in technological processes (Kemp, 1997).

Technology is a dynamic process which involves the introduction and diffusion of new technologies and as well the update and/or utilization of the existing technologies. The dynamism of technology determines or accounts for technological trajectories which are important factors in achieving social, economical and environmental goals at both local and global level. Hence, the mechanisms and driving forces of the technology changes, as well as characteristics and structure of both prevailing and emerging regimes (technologies) must be understood in order to conceive the actions necessary to stimulate the change in the required direction (Kemp, 1997).

Today, it has become apparent that the critical issue for industrial countries and firms is not the technology they have at present, but rather, how they learn to evaluate, adopt, and use the technology effectively to improve their products and services. What endow a competitive edge on a firm or a nation is what it knows, how it uses what it knows and how fast it can know something new (Prusak, 1996). With the pace of industrial transformation in developing countries, firms are experiencing a growing need for innovation to remain relevant in the market place (Kemp, 1997). Unfortunately, the absence of strong technological capability which contributes to sustainable competitiveness often impedes the firms from improving their innovations (Prusak, 1996).

In Nigeria and other developing countries, innovation is seen as a process of adoption, absorption and diffusion of traditional available technology (Lall, 1991). It is thus important for firms to develop the capability to exploit the locally available technology as a means of gaining competitive edge. This capability is embedded in the set of skills and competences which the firm possesses and/or can exploit to manage change (Oyelaran-Oyeyinka *et al.*, 1996). Innovation capability consists of the skills, knowledge, resources, creativity and attitude needed to continuously generate and translate ideas and change into marketable new or significantly improved devices, products, processes and services (Tidd, 2001). Innovation is very essential for the growth of firms and also for the firms to sustain competitive advantage (Souitaris, 2003; Hitt *et al.*, 1997; Tidd, 2001). Innovation has been strongly linked to economic growth (Tidd, 2001). As a nation's economic growth may be understood by a lay man as the nation's increasing capacity, to produce more and better goods and services to its population. Technological innovation is defined as the first commercial application of a new technology which occurs when a new or changed product is introduced to the market, or when a new or changed process is used in commercial production (Souitaris, 2003). The emergence of a new technology, which may take the form of a product, process or service, is a result of several activities spanning an appreciable length of time depending on the type of technology (Hitt *et al.*, 1997).

Industrialization has led to global competition among nations and firms which in turn has caused firms to innovate over time. A firm that intends to perform today and cope with the future must therefore incorporate innovation into its organizational routine. But this has not really been the case (Storey, 2000). Innovation in firms has been influenced more by economic development and the management culture rather than competition (OECD, 2005; Oyelaran-Oyeyinka *et al*, 1996). Owing to these, and some other reasons such as systemic weaknesses and weak research-industry interaction (Ilori *et al.*, 2000; Oyewale, 2005; Oyebisi *et al.*, 1996), a sharp contrast exists between firms in developing nations and the developed ones. Hence, innovation studies in technologically advanced countries cannot be correlat ed with innovation patterns in the third world countries.

Furthermore, the main factors promoting innovation in firms especially in the developing countries of the world is important for policy. In addition to innovation activities, a number of other factors may affect technological learning, capability and innovation (OECD, 2005). While there has been much study on the factors that influence firms' propensity to innovate, little of this has been in developing countries especially in Africa. This study is premised on the need to close this gap in development-relevant knowledge in Nigeria.

2. Research Questions

This study seeks to answer the following questions through empirical evidence:

- What types of technological innovation had occurred in the indigenous oil and gas servicing firms in Nigeria?
- How intense is innovation in this sub-sector?
- What are the specific factors that influence the capability of these firms to innovate?

3. Methodology

The main analytical concepts and the relationships in this study is based on established relationships in the literature on technological capabilities and represents a modification on the one applied by Romijn and Albaladejo (2002) in their study of the determinants of innovation capability in small high-tech firms in the United Kingdom (UK) and Egbetokun *et al,* (2007) in their study of the assessment of innovation capability in the cable manufacturing industry in Nigeria. Capabilities accumulate as a result of various internal and external inputs (Romijn and Albaladejo, 2002). Although Romijn and Albaladejo's (2002) results showed that not all of the inputs in the model are significantly related to innovation capability, they have all been included in this assessment for two main reasons. That is difference in industry and country contexts as well as the larger number of innovation types considered in this study.

This study focused on Nigeria's indigenous oil and gas servicing firms. The firms were purposefully selected. Data was collected from the firms through the use of a structured questionnaire. The questionnaires were administered on four departments identified as relevant to this study (Production, Engineering/Maintenance, Finance and Administration Departments). Sixty questionnaires were administered to the firms but only forty which were completed with accuracy was used in the analysis.

The questionnaire elicited information on the innovation strategies of the firm, its human resource, innovation rate and expenditure on R&D as well as the incidence of collaboration/ joint activities with stakeholders within the Petroleum Innovation System (PIS). The questionnaire, which is for the Production Department and Engineering Department, covers the innovation and technological efforts of the firms; their sources of information for innovation as well as the nature of linkages and interactions in the industry.

Proxy variables of the firm's innovative performance as reflected in whether it has innovated and the type(s) of innovation it has undertaken (UNU-INTECH, 2004) were used in the study. To measure the technological innovation, this study considered four types of innovation as outlined:

- Product innovation was measured via three proxy variables, one of which is similar to those used in many previous studies (Romijn and Albaladejo, 2002). The first variable indicated whether or not the firm developed a new product during the reference period. The next two variables measured whether or not the firm introduced a new product to the market or modified an existing one.
- Process innovation was measured via three proxy variables indicating whether the firm developed, introduced a new process in its production activities or modified an existing one (Romijn and Albaladejo, 2002; UNU-INTECH, 2004).
- Organisational innovation was captured with four variables drawn from UNU-INTECH (2004): changes in quality controls; maintenance routines, waste management procedures and reverse engineered any product or process
- Diffusion-based innovation is considered to be unique in the sense that it is particularly important in the developing country context. This innovation type was measured through two

simple binary variables: the acquisition of product licence and/or process licence. Whether or not a firm uses technology licensed from a foreign owned company or enters into a joint venture has been found to be important for firm growth.

UNU-INTECH (2004) noted that the propensity to innovate is related to:

- Location of the firm within or close to a major urban area;
- Proximity to sources of new knowledge and ease in participating in knowledge flows;
- Educational level of the Owner/CEO/Manager, especially a degree from a technical university or engineering programme that stimulates and facilitates problem solving;
- Global exposure through training, work or study abroad which opens opportunities for networking for knowledge flows and collaboration and creates an awareness of the utility to do so;
- Ownership structure of the firm, which influences the choice of products and processes as well as their subsequent modification or change. For instance, technology within a multinational group is often sourced from the foreign parent company and thus, the effect of being an affiliate of a foreign firm should be accounted for;
- The firm's sector, which provides a measure of the stimulus to innovation resulting from the higher R&D intensity of the sector and nature of competition within the sector;
- The size of the firm, which is related to its access to resources and opportunities for knowledge scanning to support a process of innovation. Among firms, small enterprises are better suited to carry out technological innovations due to their specific advantages of flexibility, concentration and internal communications ;
- Exports (as a percentage of sales) and whether this is rising as an indicator of the firm's competitive interests and abilities; and
- Habits and practices of innovation as reflected in having innovated in the past.

Five key areas of influence on an organisation's ability to innovate are generally identified in the body of literature which was reviewed for this study, and these can clearly be categorised, relative to the firm, into internal and external components. These influences relate to leadership, opportunistic behaviour, culture and change, learning, and networking and relationship building.

Internal Sources of Innovation Capability

- The professional background of the heads of Technical Departments was captured through three variables. The educational qualification was represented by multiple-choice measures of management, science and engineering and other academic degrees obtained. Prior work experience abroad was represented as the number of years that the heads of technical departments worked in any of small enterprises, large corporations, and university or related institutions. Prior training experience abroad in small enterprises, large corporations, and university or related institutions to current work were measured on a Likert scale.
- Human Resources of the firms were measured by variables representing the extent of organisational learning and the working proportion of R&D staff in total workforce was also considered (Romijn and Albaladejo, 2002).
- Investment effort was captured by variables representing innovation, R&D and training investment (Romijn and Albaladejo, 2002).
- External Sources of Innovation Capability
- Networking and collaboration was captured by the incidence of contacts with external agents. For example, relationships with customers, suppliers, enterprises in related lines of business, financial institutions, universities, research institutions, training institutions, service providers and industry.
- Proximity advantages from networks measured the geographical proximity advantages associated with the above network interactions and these were measured by asking the respondents to indicate, whether or not a proximity advantage was attached to each of the interactions

- Institutional support measured by whether or not firms had received financial support from government or other forms of support from other institutions such as trade associations. The information gathered was supplemented with additional information from published sources [e.g. Nigerian National Petroleum Corporation (NNPC), Department of Petroleum Resources (DPR)] and in-depth interviews, as well as field observations. Previously tested questionnaires such as the African template proposed by UNU-INTECH (2004), the revised questionnaire employed in the South African Innovation Survey 2001 (Oerlemans *et al.*, 2001) and the list of variables used by Romijn and Albaladejo (2002) and Egbetokun *et al.* (2007) in their studies on the determinants of innovation capability in UK SMEs and Nigeria cable manufacturing industry were consulted in designing the questionnaire for this study.

4. Results and Discussion

Types of Technological Innovations

The study considered the prevalence of different types of innovative activities in the sub-sector. Table 1 shows the statistics on the prevalence of each types of technological innovations found within the indigenous oilfield servicing sub-sector. It shows that not less than 30% of the firms had engaged in product, process and organisational innovations during the 2001–2010 periods. On the average,29% of the firms were innovation-active during the reference period. This is lower than the innovation prevalence of 52.7%, in a study of small enterprises in the state of Karnataka, India (Bala-Subrahmanya, 2005) and also lowers than the innovation prevalence72% of the firms in the manufacturing sector (NACETEM, 2010). Certain anomalies become obvious when one takes a closer look at the figures in Table 1.

Table 1: Showing the Types of Technological Innovation that had taken Place between 2001 and 2010

	Types of Technological Innovation	Yes (%)	Started but later abandoned (%)	No (%)
Product Innovation	Introduced new product	12.5	2.5	85
	Developed new product	2.6	5.1	92.3
	Improved existing product	5.1	0	94.9
Process Innovation	Introduced new process	12.8	5.1	82.1
	Developed new process	2.6	10.3	87.2
	Improved existing process	23.1	5.1	71.8
Organisational Innovation	Introduced quality Control	42.5	7.5	50
	Introduced maintenance routine	55	10	35
	Introduced waste management procedure	37.5	15	47.5
	Reverse engineered any product/process	10.3	28.2	61.5

Diffusion-Based Innovation

Of the total, 41% (16) of the firms developed/applied any new or modified versions of any of either product or process technology within the period 2001 and 2010. About 31% (5) firms developed the new or modified versions of product or process which they have a license on their own while 62.5% (10) of the firms developed or modified through partnership with the licensor.

The prevalence of organisational innovations was observed to be higher than that of all other innovation types. It was the only type of innovation that ranked high. This primarily suggests that organisational changes are at the heart of the innovation processes in the indigenous oilfield servicing sub-sector. These changes, like OECD (2005) argued, are typically expressed in business practices and workplace organisation that are new to the firm and occur as a result of strategic management decisions. The intensive prevalence of organisational innovation, alongside process innovation, within our developing country context is not surprising because organisational changes, very much like process innovations, are less risky and consume much less resources compared to other types of innovations. Concerning process innovation, Egbetokun *et al.* (2009) noted that changes in processes are less rigid, more responsive to 'shop-floor' serendipitous discoveries and may not generally require financial investments as much as product innovations.

The 41% incidence of diffusion-based innovation recorded in this study actually arose from product and process licences. This somewhat low figure is a true representation of the situation within the sub-sector. The firms in the industry are all SMEs that are mostly confronted by several resource constraints. It is therefore difficult for them to procure licences. Low levels of absorptive capacities may also explain this situation (Audretsch *et al*, 2005; Rosa and Mohnen, 2008). The firms that have been able to do this are those which are part of a Foreign-based corporation

Intensity of Technological Innovations

Training Expenditure

Table 2: Showing the Amount Range Expended on Staff Training within the Period 2007-2010

Amount (Million Naira)	2007 (%)	2008 (%)	2009 (%)	2010 (%)
Less than 1	66.7	50	35	30.5
Between 1 and 5	33.3	50	65	60.8
Between 5 and 10	Nil	Nil	Nil	8.7

The training expenditure of the firms is given in Table 2. Majority of the firms that engaged in staff training spent below a million naira per year but the trend became reversed as the years progressed (see Table 2)

Innovation Expenditure

Table 3: Showing the Budget Range Expended on Innovation within the Period 2007-2010

Amount (Million Naira)	2007 (%)	2008 (%)	2009 (%)	2010 (%)
Less than 1	55.5	50	50	41.7
Between 1 and 5	44.5	50	50	50
Between 5 and 10	Nil	Nil	Nil	8.3

The innovation expenditure of the firms is given in Table 3. Like the training expenditure, the innovation expenditure increased as the years progressed (see Table 3)

Internal Factors that Affect Innovation Capability of the Firms

The main internal sources of innovation capability evaluated in this study are the background of the head(s) of technical departments in the firm and the internal technological efforts of the firm (staff training and innovation expenditure).

Background of the head(s) of technical department(s)

11 out 40 have separate production and engineering departments while the remaining 29 have the production and engineering department as one single department.

Production department

52.5 % of the heads of production department have Higher National Diploma, while 20% have a Bachelors Degree, 22.5% have Masters Degree and 5% have Doctorate Degree. 80% of the heads of production department have degrees in Science or Engineering field while 20% have degrees in Management or Finance-related field. 7.5% of the heads of technical department went for training in other African countries 7.5% in Europe, 20% in Europe, 7.5% in South America, 15% in Asia and 5% in Australia at some point in time. 2.5% of the head of the production department obtained a degree in other African Countries, 15% obtained degree in Europe, and 2.5% obtained degree in United States, Canada or Mexico.

Engineering department

12.5% of the heads of engineering department have Higher National Diploma, while 30% have Bachelors Degree, 20% have Masters Degree and 7.5% have Doctorate Degree. 77.5% of the degree is in Finance-related field. 15% of the heads of engineering department went for training in other African Countries, 12.5% in Europe, 15% in United States, Canada or Mexico, 5% in South America, 12.5% in Asia and 2.5% in Australia at some point in time.10% of the heads of the engineering department obtained a degree in Europe, 2.5% in United States, Canada or

Mexico while 2.5% of the heads of the engineering department had previously worked in other African Countries.

Internal Technological Factors

Training and technological learning

Table 4 shows the number of firms that had training (learnt some new technologies) within the period 2007 to 2010.

Table 4: Showing the Frequency of Training that took Place in the Firm within the Period 2007 to 2010

	2007	2008	2009	2010
Use of Information & Communication Technology	3 (7.5%)	4 (10%)	7 (17.5%)	7 (17.5%)
Repairs and Maintenance	1 (2.5%)	5 (12.5%)	12 (30%)	10 (25%)
Project Management	1 (2.5%)	2 (5%)	12 (30%)	14 (35%)
Technical Report Writing	0	1 (2.5%)	4 (10%)	2 (5%)
Industrial Safety	2 (5%)	4 (10%)	8 (210%)	8 (20%)

The fact that firms require an adequate stock of skilled manpower and the role played by firm-level investments in training in enhancing this pool has been established in the innovation literature (Romijn and Albaladejo, 2002; Amara *et al.*, 2008). More recent research has indeed proven that firms that continually invest in staff training tend to be more capable to innovate. For instance, Amara *et al.* (2008) found that Canadian manufacturing SMEs that carried out staff training activities were 1.92 times more likely to innovate than their counterparts who did not. The results in Table four showed that that the indigenous oilfield servicing firms in Nigeria were marginally active in this regard as all of the firms reported having implemented one or more staff training programmes within the period covered in this study

Research and development efforts

41% (16) of the firms developed/applied any new or modified versions of any of either product or process technology within the period 2001 and 2010. 31.2% (5) firms developed the new or modified versions of product or process which they have a license on their own while 62.5% (10) of the firms developed or modified through partnership with the licensor.

Research and development activities

37.5% of the firms are involved in research and development activities but 15.4% of the firms have separate research and development department. In 2007 and 2008, there were a total of twenty-four personnel working in the research and development department in five out of the forty firms that were selected for this study. In 2009, the number rose to thirty-seven and in 2010, six firms now have separate research and development department with a total number of forty-six personnel.

External Factors that Affect Innovation Capability of the Firms

Interactions within the petroleum nnovation system

The study considered networking and collaboration as well as the advantages that firms might derive from these when they are located close to the actors involved in these networks. Several previous studies (Meeus *et al.*, 1999a, 1999b; Romijn and Albaldejo, 2002) have highlighted the importance of a number of stakeholders within an innovation system that firms may network or collaborate with. Evidences exist in favour of customers, suppliers, industry associations, higher education and research institutions, among others as helpful sources of information for the firms' innovation activities.

Table 5 shows that the firms interacted more with Industry Association than with most other actors. Egbetokun *et al.* (2009) found that the most highly rated source of information for innovation were customers. Jaruzelski and Dehoff (2007) also showed that customers probably matter more than any stakeholder in innovation. Interaction with competitors and Training Institutions were also relatively intense as 65% of the sampled firms collaborated with each of these stakeholders during the period covered by this study.

All the firms are subcontracting for the Oil and Gas companies. 74.4% of the oil and gas servicing firms outsource to other oil and gas servicing firms.

Table 5: Intensity of the interactions between the elements of Nigeria's Petroleum National Innovation System

	(%)
Competitors	75.5
Customers	61.8
Suppliers	64.7
Associated Companies	40.6
Consulting and Marketing firms	21.9
Private Research Institutes	16.7
Public Research Institutes	13.3
Universities	36.4
Government Ministries	6.9
Financial Institutions	29
Training Institutions	67.6
Industry Association	92.1

In contrast with the general trend in the literature (e.g. Autant-Bernard and Massard, 2001; Keller, 2002; UNU-INTECH, 2004; Audretsch *et al.*, 2005; Rosa and Mohnen, 2008), the occurrence of proximity advantages from knowledge centres among the firms was very low except for the Training Institutions (see Table 5). This is in spite of the fact that all the firms are located in the same city with at least a university. Perhaps this is explained by the general economic paradigm that firms require a certain level of absorptive capacities (usually approximated by firms' proportion of R&D staff, engineers or scientists) to be able to assimilate scientific knowledge and to benefit optimally from partnerships with knowledge centres, especially universities (Cohen and Levinthal, 1990; Fontana *et al.*, 2006; Rosa and Mohnen, 2008). Although it was difficult to empirically evaluate the firms' absorptive capacity because data gathered on staff profiles were largely partial and unreliable, the information gathered from the interviews conducted pointed out that the absorptive capacities of the firms in the indigenous oilfield servicing sub-sector was quite low. For instance, in few of the firms, apart from the CEO and his heads of technical departments, every other staff had less than a university degree. Within the production department in those firms, nearly every worker was employed with secondary or vocational qualifications and then trained on the shop floor. In some firms, we found about 75% of employees having either secondary school or vocationalqualifications. Proportion of engineers and scientists in firm's total workforce was 15% while the proportion of technicians was 10%.

Table 6: Advantages from the proximity between the elements of Nigeria's Petroleum National Innovation System

	(%)
Competitors	80.6
Customers	65.4
Suppliers	70.4
Associated Companies	40
Consulting and Marketing firms	13.6
Private Research Institutes	13.6
Public Research Institutes	18.2
Universities	38.5
Government Ministries	5
Financial Institutions	28
Training Institutions	69
Industry Association	94.3

On the other hand, majority of the firms admitted that close location to suppliers (70.4%), competitors (80.6%), customers (65.4%), training institutions (69%) and particularly industry associations (94.3%) had been very beneficial to them (see Table 6). This seems to suggest that it could have been easier for the firms to pursue and seize the advantages from being close to these actors since no considerable level of absorptive capacity is required. An implication that could be drawn from the foregoing discussion is that in facilitating industry-academic relations, both proximity and firm-level absorptive capacities are critical and require attention from all stakeholders.

Government Support

Previous studies have shown that major government S&T policies and programmes may have more impact on innovation than the activities and strategies of private enterprises (OECD, 2005). Thus, the role of government as an institution is critical for firm-level innovation. Such roles typically include the design and implementation of innovation-friendly policies, effective monitoring of these policies, procuring innovative products from domestic firms and creating a stable political and economic ambience, among others. 2.6% of the selected firms enjoyed support from the Nigerian government within the period 2001 to 2010 (see Table 7) while the remaining 97.4% did benefit an form of government support within the said period though 33.3% of the selected firms claim to be aware of some government policies that support innovation.

Perhaps even more striking is the fact that the firms have apparently not collaborated extensively with or gained serious proximity advantages from government organisation (public research institutes and government ministries). Only 6.9% of the firms have collaborated with government ministries and 5% indicated gaining proximity advantages from these, 13.3% have collaborated with public research institutes while 18.3% indicated gaining advantages from these. While the latter is understandable as the firms are not so closely located to the ministries, the dearth of collaborative activities with them seems to suggest that these institutions have not been adequately supportive.

Table 7: Advantages the Selected Firms have Benefited from Government

	R&D funding (%)	**Training (%)**	**Subsidies (%)**	**Tax Rebate (%)**	**Technical Support (%)**	**Loans and Grants (%)**
Not Important	7.5	5	0	0	2.5	10
Slightly Important	7.5	10	5	0	7.5	2.5
Important	0	0	7.5	17.5	5	5
Very Important	5	5	20	20	12.5	2.5

Table 7 reflects that the firms have not really benefited from government support and policies in the study period. There is need for government to formulated new policies that would support the growth of small and medium scale enterprises.

5. Conclusion

The results showed that these firms, which are mostly SMEs, demonstrated some level of innovation capabilities, albeit in an uneven manner. Although some product, process and diffusion-based innovation were found, organisational innovations were at the heart of the innovation activities of the firms. In addition, it was found that these firms operate within a weak Petroleum Innovation System (PIS). It is to be concluded from these that firms operating within such contexts are not necessarily innovation-inactive. However, they might not be able to engage in implementing product and process changes that require much knowledge and financial investments. Given the generally deficient state of the innovation influencers, it is not surprising that the firms were more capable to implement changes in organisation and processes than what they would do with their products and embodied knowledge from elsewhere.

The importance of firm-level investment in organisational learning is further brought to the fore by the findings. Firms are required to consciously make investments in developing a capable stock of human capital. When this happens within a network of strong institutions and a favourable economic ambience, a firm will definitely be much more capable for innovation. This is to say that while the firms make efforts to build capabilities, the success of their efforts depend directly on the supportiveness of the environmental context within which they operate. An importantly new knowledge advanced by this study is the specific role that industry associations could play in facilitating firm-level innovation capability. It has been indicated by the results that industry associations have a central role to play in facilitating firm-level innovation especially in the developing country context. Few earlier studies have identified such associations as important in the processes that lead to the build-up of firm-level capabilities and the knowledge available on the specific roles that they are capable of playing is still very sparse. Obviously, their roles now transcend mere activism and protection of rights but also involve covering resource deficiencies for member firms, helping member firms learn and creating access to innovation-friendly support.

The need for a strong innovation system is also indicated by the findings of this study. Firms were seen not to have received much support from knowledge and centres and even from government. Specifically, diffusion-based innovation was very low. The few firms that succeeded in doing this were those that on their own had significant external resource endowments by virtue of belonging to a global group or creating international ties. Most of the firms were largely unable to muster enough resources on their own to engage in activities that would give rise to that kind of innovation. Thus, stronger government-finance-research-industry linkages that would ease resource deficiencies are critical to firm-level innovativeness.

References

Abereijo, I. O., Ilori, M. O., Taiwo, K. A. and Adegbite, S. A. (2007), 'Assessment of the capabilities for innovation by small and medium industry in Nigeria', in *African Journal of Business Management*, 1(8): 209-217.

Adjebeng-Asem, S. (1996), 'Translating Technical Innovations into Entrepreneurship in Nigeria: Social and Policy Implications', In: Ogbu, O. M., Oyelaran-Oyeyinka, B. and Mlawa, H. M. (eds.), *Technology Policy and Practice in Africa,* Ontario: International Development Research Centre. Available at: http://www.idrc.ca/en/ev-30793-201-1-DO_TOPIC.html (Accessed: March, 2011).

Audretsch, D.B., Lehmann E.E. and Warning, S. (2005), 'University Spillovers and new Firm Location', in *Research Policy,* 34(7): 1113-1122.

Autant-Bernard, C. and Massard, N. (2001), 'The Geography of Knowledge Spillovers and Technological Proximity', In: Feldman, M. and Massard, N., *Institutions and Systems in the Geography of Innovation,* The Netherlands: Kluwer Academic Publishers Group.

Bala-Subrahmanya, M. H. (2005), 'Technological innovations in Indian small enterprises: dimensions, intensity and implications', in *International Journal Technology Management,* 30(1/2): 188-204.

B.P. Statistical Energy Survey (2008), Available at: http://www.mbendi.com/indy/oilg/af/ng/p0005.htm (Accessed: 17 January 2011).

Chesbrough, H. (2003), *Open Innovation: The new Imperative for creating and profiting form Technology*, Boston, MA: Harvard Business School publishers.

Cohen, W.M. and Levinthal, D.A. (1990),'Absorptive Capacity: A New Perspective on Learning and Innovation', in *Administrative Science Quarterly*35: 128-152.

Drucker, P. F. (1985), *Innovation and Entrepreneurship: Practice and Principles*, New York: Harper and Row Publishers.

Egbetokun, A. A., Siyanbola, W. O. and Adeniyi, A. A. (2007), 'Assessment of Innovation Capability in the Cable and Wire Manufacturing Industry in Nigeria: a case study approach', *Paper presented at the Micro Evidence on Innovation in developing Economies[MEIDE]*, UNU-MERIT, Maastricht, the Netherlands, May 31-June 1, 2007. Available at: http://www.merit.unu.edu/MEIDE/papers/2007/EGBETOKUN_SIYANBOLA_ADENIYI_Assessment%20of%20innovation%20capability... cable%20and%20wire%20manufacturing%20industry.pdf (Accessed: 11 January 2011).

Egbetokun, A.A., Siyanbola, W.O., Sanni, M., Olamade, O.O., Adeniyi, A.A. and Irefin, I.A. (2009), 'What drives innovation? Inferences from an industrywide survey in Nigeria', in *International Journal of Technology Management*, 45(1/2): 123-140.

Fontana, R., Geuna, A. and Matt, M. (2006), 'Factors affecting university-industry R&D projects: The importance of searching, screening and signalling', in *Research Policy*,35(2): 309-323

Ilori, M.O., Oke, J. S. and Sanni, S. A. (2000), 'Management of new product development in selected food companies in Nigeria, in *Technovation*, 20(6): 333-342.

Jaruzelski, B. and Dehoff, K. (2007), 'The Customer Connection: The Global Innovation 1000', from *Strategy+Business*, Issue 49, Winter 2007,Winter, Booz & Company. Available at: http://www.urenio.org/wp content/uploads/2009/01/global-innovation-1000-2007.pdf

Kemp (1997), *Imitation to Innovation: The Dynamics of Korea's Technological Learning,* Boston, MA.: Harvard Business School Press.

Keller, W. (2002), 'Geographic Localization of International Technology Diffusion', in *American Economic Review*92(1): 120-142.

Lall, S., (1991), 'Human Resources, Technology and Industrial Development in Sub-Saharan Africa', In: Chhiber, A. and Stanely, F. (eds.), *Economic Reform in Sub-Saharan Africa,* Washington D.C.: World Bank.

Meeus, Marius T.H., Oerlemans, Leon A.G. and Van, D., and Jules, J.J. (1999a), 'Regional systems of innovation from within: An empirical specification of the relation between technological dynamics and interaction between multiple actors in a Dutch region', *Working Paper 99.1, ECIS,* Eindhoven Centre for Innovation Studies, The Netherlands.

Meeus, Marius T.H., Oerlemans, Leon A.G. and Hage, Jerald (1999b), 'Sectoral Patterns of Interactive Learning: An empirical exploration using an extended resource based model', *Working Paper 99.5, ECIS,* Eindhoven Centre for Innovation Studies, The Netherlands.

Oerlemans, L.A.G., Meeus, M.T.H. and Boekema, F.W.M. (1998), 'Learning innovation and proximity', *Working Paper 98/3, ECIS,* Eindhoven Centre for Innovation Studies, Eindhoven University of Technology, The Netherlands.

Organisation for Economic Cooperation and Development (OECD) (2005), 'Proposed guidelines for collecting and interpreting technological innovation data: Oslo Manual', 3rd edition, Paris: OECD.

Oyebisi, T.O., Ilori, M.O. and. Nassar, M.L. (1996), 'Industry-academic relations: an assessment of the linkages between a university and some enterprises in Nigeria', in *Technovation,* 16(4): 203-209.

Oyelaran-Oyeyinka, B. (1997), *NNEWI: An Emergent Industrial Cluster in Nigeria,* Ibadan Technopol Publishers.

Oyelaran-Oyeyinka, B. and Lal, K. (2006), *SMEs and New Technologies, Learning, E-Business and Development,* Basingstoke, UK: Palgrave Publishers.

Oyelaran-Oyeyinka, B. (2001), 'Networks and Linkages in African Manufacturing Cluster: A Nigerian Case Study', DP 2001-5, September, UNU-INTECH, Maastricht, The Netherlands

Oyelaran-Oyeyinka, B. (2002), 'Manufacturing Response in a National System of Innovation: Evidence from the Brewing Firms in Nigeria', *DP 2002-3,* April, UNU-INTECH, Maastricht, The Netherlands.

Oyelaran-Oyeyinka, B. (2004), 'Learning and Local Knowledge Institutions in African Industry', *DP 2004-2,* UNU-INTECH, Maastricht, The Netherlands.

Oyelaran-Oyeyinka, B., Laditan, G. O. A. and Esubiyi, A. O. (1996), 'Industrial innovation in Sub-Saharan Africa: the manufacturing sector in Nigeria', in *Research Policy* 25(7): 1081-1096.

Oyelaran-Oyeyinka, B. and Barclay, L. A. (2004), 'Human Capital and Systems of Innovation', in *African Development Review,* 6(1): 115-138.

Oyelaran-Oyeyinka, B. (2003), 'Innovation and Learning by Firms in Nigeria: The Role of Size, Skills and Ownership', in *International Journal of Business and Society,* 4(1): 1-22.

Oyelaran-Oyeyinka, B. (2003), 'Knowledge Networks and Technological Capabilities in African Clusters', in *Science, Technology and Society,* 8(1): 1-24.

Oyelaran-Oyeyinka, B. (2005), 'Inter-Firm Collaboration and Competitive Pressures: SME Footwear Clusters in Nigeria', in *Int. J. Technology and Globalization,* 1 (3/4): 343-360.

Oyelaran-Oyeyinka, B. (2006), 'Systems of Innovation and Underdevelopment: An Institutional Perspective', in *Science, Technology and Society,* 11(2): 239-269.

Oyelaran-Oyeyinka, B., Adelaja M. and Abiola, B., (2007), *SMEs Clusters in Nigeria: Institutions, Policies and Support Systems,* Ibadan: Booksellers Publishers.

Oyelaran-Oyeyinka, B. (1997), 'Technological Learning in African Industry: A study of Engineering Firms in Nigeria', in *Science and Public Policy,* 24(5): 309-318.

Oyewale, A. A. (2005), 'Addressing the research-industry linkage impasse in Nigeria: The critical issues and implementation strategies', *Paper presented at The 3rd Globelics International Conference*, Tshwane University of Technology, Pretoria, South Africa, 2005. Available at: www.globelics2005africa.org.za/papers/p0032/Globelics2005_Adesina%20Ayobami%20Oyewale.pdf (Accessed December 2, 2010).

Oyewale, A.A. (2003), 'Evaluation of the Interactions Among the Key Components of Science and Technology and Innovation System in Nigeria', *PhD Thesis*, Technology Planning and Development Unit, Faculty of Technology, Obafemi Awolowo University, Ile-Ife, Nigeria.

Ozigbo N. C. (2008), 'Technological Capacity Building in the Nigeria's Oil and Gas Industry', *Proceedings of the 19th Annual Conference of the International Information Management Association*, San Diego, California.

Prusak L. (1996), 'The knowledge advantage', in *Strategy and Leadership*, 24(March-April): 6-8.

Romijn, H. and Albaladejo, M. (2002), 'Determinants of innovation capability in small electronics and software firms in southeast England', in *Research Policy*, 31: 1053–1067

Rosa, J. and Mohnen, M. (2008), 'Knowledge Transfers between Canadian Business Enterprises and Universities: Does Distance Matter?' *UNU-MERIT Working Paper #* 2008-17. Available at: http://www.merit.unu.edu/publications/wppdf/2008/wp2008-017.pdf (Accessed March 21, 2011).

Storey, J. (2000), 'The management of innovation problem', in *International Journal of Innovation Management*, 4(3): 347-369.

Souitaris, V. (2003), 'Determinants of technological innovation: Current state of the art, modern research trends and future prospects', In: Shavinina L. V. (ed.), *The International Handbook on Innovation*, Oxford: Elsevier Science.

Tidd, J., Bessant, J., and Pavitt, K. (1998), 'Managing innovation: Integrating Technological, Market and Organizational Change', 2nd edition, Chichester, UK: John Wiley and Sons.

Tidd, J., Bessant, J., and Pavitt, K. (2005), 'Managing Innovation: Integrating Technological, Market, and Organizational Change', 3rd edition, Chichester, UK: John Wiley and Sons.

Tidd, J. (2001), 'Innovation Management in context: Environment, organization and performance', in *International of Mnangement Reviews Journal,* 3(3): 169-184.

United Nations University Institute for New Technologies (UNU-INTECH) (2004), 'Designing a Policy-Relevant Innovation Survey for NEPAD', mimeo: A Study prepared by UNU-INTECH, Maastricht, The Netherlands.

Chapter 12

Financial Lending and Investment in Sustainable Small-Scale Agribusiness in Sub-Saharan Africa: A Review of Carbon Sequestration and Welfare Benefits

Benjamin Olatunbosun

1. Introduction

Agricultural growth plays a critical role in eradicating poverty especially in low-income countries. In developing countries, especially sub-Saharan Africa, apart from generating a substantial level of revenue, the sector contributes ca. 32 per cent to Gross Domestic Product (GDP) (Christiaensen and

Demery, 2007). About 75 per cent of the work-force in low income countries in sub-Saharan Africa is engaged in agriculture-with largely small-scale farming of between 0.5 and 2 hectares of land (The World Bank, 2007). In promoting agricultural growth, it is essential to increase factors of production and productivity (labor – L, capital – K and total factor productivity – A) in a sustainable manner as this ensures food and energy security. The ability to alleviate poverty depends to a large extent on the availability of finance as both urban and rural business activities must have access to regular and adequate capital as well as resource management knowledge to function (Balkenhol, 1991). Finance and Capital to rural small-scale agribusiness requires well-defined and stable financial institutions as the investment and individual wealth or asset of rural smallholders are quite digress and the absence of these institutions may lead to a plunge in capital stock during economic downturn (Gonzalez-Vega and Graham, 1995).

The composition of rural financial institutions has evolved from Micro Finance Institutions (MFIs), money lenders and Non-Governmental Organizations (NGOs) to include commercial banks due to the gap in the access to wide range of credit (Randhawa and Gallardo, 2003; Cull *et al.*, 2008). The prime objective of micro finance remains the ability to providing financial services to low-income households, micro, small and medium size firms at affordable rate to support economic growth while utilizing appropriate methodology that guarantees financial soundness and enhance consumer livelihood. Sub-Saharan pre-regulatory eras of micro finance in the 1980s were characterized by structural disorganization, high default rate and lack of financial discipline. In order to foster a stable financial environment international and national prudential organs have stipulated guidelines or regulations which financial institutions specifically deposit-taking organizations have to abide by (Amha, 2001). As financial and non-financial institutions in the sub-Saharan Africa regions are increasing regional operations, it is important that these institutions also adhere to international standards such as the Basel accord. Stakeholders and depositors protection as well as monitoring and measuring of risk is a vital component these guidelines (Lafourcade *et al.*, 2005). loans to off-farm enterprise has shown remarkable success due to reduced vulnerability, stable income and group guarantee, however, micro financing of agriculture is not as profitable as conventional lending with collateral or collateral substitute (Cull *et al.*, 2008).

An array of literatures identifies credit constraints as one of the major problems confronting development of agriculture not only in developing countries but on a global scale. Factors ranging from information asymmetry, moral hazard, contract enforcement, high rate of interest and farm debt, under-developed financial institution to high transaction cost may be responsible for limited credit allocation.In the 1970s, 15 per cent of farmers had access to institutional credit in Latin America and 5 per cent Africa (Gonzalez-Vega, 1993; Gonzalez-Vega and Graham, 1995). The issue of land tenure system has compounded to the woes of agricultural financing because lack of proper definition of right to land deprives usage as collateral while the absence of stable income arising from risk associate with agriculture (weather, price volatility, poor infrastructure) means no physical collateral. The direct effect of climate change on the banking operations can be due to severe weather threat or indirectly due to regulatory compliance or change in societal preference.Financial institutions in rural areas can, however, seize business opportunities to provide services and develop financial techniques that foster climate mitigation and adaptation through emission trading and advisory services (Furrer *et al.*, 2009).

Africa will be one of the most vulnerable regions to the effect of climate change, due to projected decrease in precipitation and increase in temperature and population as well as decline in GDP (Schneider *et al.*, 2007). Therefore there is a need to shift attention to what can be done not only to adapt but also to mitigate its negative in order to ensure food security for the growing population. Investment in adaptation and mitigation of harmful climate change (including agriculture – carbon sink) requires an estimated investment of over US$1 trillion per annum with 86 per cent anticipated to come from the private sector, such as financial institutions, directed to developing countries. Although the private sector seems discouraged by carbon market situation, there are certain investments that are starting to emerge (Jones and Hiller, 2010). Climate smart agriculture can be propelled by smallholders and may provide sustainable investment potential of ca. US$1 trillion over the next 4 decades which has been projected for sub-Saharan Africa (Branca, 2011). There are signs of strong interest in investing in sustainability amongst financial investors on a global scale. A study by the Rockefeller foundation found that 69 per cent of financial advisory perceived sustainable investment as an avenue for growth.However, only 38 per cent actively participated in international micro-finance engaged in sustainable development while over 80 per cent were not engaged and 43 per centwere either doubtful or uninterested (Conway and Stevens, 2012).

This study seeks to establish how financial institutions engaged in microfinance can benefit directly from agricultural sustainability through climate investment or indirectly through a targeted lending with climate-backed- security, which may lead to rural development and solid financial environment. There are indications that there is need to emerge with new innovative approaches to be countered the old problems relating to credit constraint experienced by the poor at the hands of financial institutions. Specifically, this study aims to establish a benchmark that can be used for further research into climate financing.

This study proceeds by investigating the financial sector servicing the rural low-income farmers and changes in operations and compliance in order to facilitate a stable financial environment. The following section reviews the lending behaviour of financial and non-financial organisations towards smallholders and their participation in investments relating to environmental protection and development. The fourth section outlines possible benefits of sustainability to credit institutions and how this can be an integral part of their operation in line with international standards. The formulation of policies which may help prompt sustainable lending and investment are the focus of the fifth section. The study then draws a conclusion from the relevant research materials and information gathered on the topic.

2. Trends in Sub-Saharan Rural Credit Institutions

Some authors have questioned if the existence of banks in the conventional definition, in this modern times, are relevant as banks continue to engage in off-balance sheet rather than deposit taking and loan disbursement, become less dependent on core business and are exploring core competences.It is also argued on the one hand that the involvement of other non-banking organization in banking activities is undermining the *traditional banks* which are causing banks to *shrink* (Llewellyn, 1999). On the other hand financial service provided to rural smallholders in sub-Saharan African by institutions is not wide-spread as the market it is both limited and under-served thus the financial sector has continued to see growth in size and operations. A substantial number of microfinance institutions in Africa have been able to post impressive growth without compromising portfolio quality, even as competition intensifies (Fernando, 2008). There are several changes that have taken place in the financing of rural poor. However, this study would rather focus on 4 of these namely:

- Institutional expansion and profitability
- Deposit
- Regulation (Risk Management)
- Technology

Institutional Expansion and Profitability

The low entry requirement for the establishment of financial institution in Sub-Saharan Africa in the 1970sand 1980s propelled the growth of local banks and non-bank financial institutions (NBFI). This period also witnessed the setting-up of government backed developmental banks,intended for the urban and rural entrepreneurs, which proved to be inefficient, despite the decline in the number of financial institutions in the 1990s due to bank failures, commercial and other financial institutions has witnessed a revival and a more stable financial environment (Brownbridge, 1998; Steel and Andah, 2003). There has been an increase in the number of sub-Saharan financial institutions in Africa operating beyond national borders in accordance to international standards (Verhoef, 2005). However, the profit efficiency of banks in Africa is argued to be in smaller size banks than in big financial institution therefore casting doubt over the development of the continent's financial sector through the establishment of mega banks (Kiyota, 2009). This may be in line with studies which show that more unconventional financial institutions, especially non-governmental organizations (NGOs), are shifting away from non-profit towards for-profit while increasing in the number of financial institutions represented by small entity in remote areas (Campion and White, 2001; Mix 2010). The fear that a for-profit approach by micro finance will result in high market profit as well as the neglect of the poor has not materialized, instead there may be an increase in diversification in the operations by microfinance institutions in different directs (Charitonenko and Campion, 2003). This may not have a negative effect as it enables these institutions to explore new innovative ideas in financing the efficient amongst the poor. The influx of a number of commercial bank into microfinance may, to a large extent, reshape the financial landscape.The economy of scale and scope from non-microfinance activities of these institutions may be used to further develop the microfinance sector.

Deposit

Asia microfinance is a prime example of how saving (deposit) can help strengthen the productivity and increased investment, the microfinance in sub-Saharan Africa has also helped increase the number of bank account and saving holders in addition to those provided by other institutions (WSBI, 2008). On the contrary, commercial banking in Africa has rather witnessed a steady movement of asset by the wealthy to off-shore location rather than deposit to local financial institutions, which may explain the low rate of saving (The World Bank, 2006). Microfinance comprising non-bank financial institution, credit unions and financial cooperatives and a small number of NGOs reached an estimated 9 million depositors with a total volume of US$2.1 billion. The number of depositors in 2010 increased to 10.2 million clients (MIX, 2008). This trend in depositors growth is in line with a general consensus that the poor will deposit part of savings in the presence of an adequately and stable financial environment (Wisniwski, 1998). In order to protect the depositors' institutions engaged in deposit-taking are guided by standards such as those recommended in Basel's core principle for effective banking supervision (e.g. Principles 2 – permissible activities; 3 – licensing Criteria; 6 – capital adequacy; 7 – risk management procedure; 8 – credit risk, etc) (BIS, 2010).

Regulation (Risk Management)

The number of regulated sub-Saharan microfinance institutions including those operated by commercial banks in 2003 was 71, while 92 institutions were unregulated (Lafourcade *et al*, 2005). The laissez-faire approach to enacting financial regulation may not necessarily be an appropriate generalization (Arun and Murinde, 2005). In Ghana, legislation and regulation for rural and micro financial institutions are creating appropriate level of new entries and providing opportunities while improving performance. The regulator of microfinance in Ghana complies with international standards (BASEL), which focuson reserve requirements which are then changed according to a classification system based on loan recovery performance (Steel and Andah, 2003). Microfinance providers in most African countries do not only have to register but are expected to adhere to the procedures for compliance with capital adequacy and solvency requirements for deposit-taking and other institutions, these prudential rules described above as well as risk concentration limits (on single borrowers), liquidity limits, have been found to exist in other countries such as Tanzania, Benin and Guinea (Basu *et al.*, 2004). Stringent regulatory and supervision has also helped improve the microfinance in Uganda with a capital adequacy level well above the minimum requirement (UCB, 2010). Although the banking and non-banking sector in Sub-Saharan Africa has made progress in generating profit over the years, the environment in which they operate is risky due to weak legal institution and low enforcement. This high risk is argued as the driver behind high returns (Flamini *et al.*, 2009)

Technology

Technological advancement has been instrumental in advancing and increasing financial services to rural dweller. For instance, mobile phones havebeen fully integrated into the financial operation through retail payment and checking price updates in the last decades (Porteous, 2006). The use of technology in microfinance in rural areas ranging from Automated Teller Machines (ATMs) or Point of Sales (POS), Interactive voice response (IVR) technology, Phone -Internet banking, Personal digital assistant (PDA), Management information system (MIS) to Credit scoring have proven to reduce not only operating cost but also aid risk assessment (Hishigsuren, 2006). Mobile phones are gaining popularity and are used by households to remit money to family and friends in remote Sub-Saharan, leading to synergy between banks and mobile operators in countries like Kenya.It was argued that mobile banking maybe more favoured in the rural areas in a number of countries if it connects to financial institutions (Comninos *et al.*, 2008). On the operational front, the need to report to supervisors as well as keep track of business dates, develop methodology for calculating interest and lending/saving has increased the use of information and communication technology (ICT). To this end certain software over the years has been developed to specifically cater for microfinance taking their unique features into consideration (Augsburg and Schmidt, 2011).

3. Credit Institution's Agricultural Lending and Sustainable Investment in Sub-Saharan Africa

Credit constraint is a major problem confronting the development of sustainable agriculture (Feder and Umali, 1993; Fernandez-Cornejo and McBride, 2002; Ahmad, 2005). Studies have shown that by providing credit to smallholders, adoption of new technology (e.g. hybrid maize) is encouraged and the ability of smallholders to bear risk increases (Diagne *et al*, 2009). All studies found that a credit constraint had a negative impact on the adoption of agricultural innovation, which ultimately might lead to limited agricultural growth and development and increased poverty. A number of rural smallholders source their savings for funds (when and if available) or from family and friends or other sources such as rich people or money lenders in the community who usually charge above market interest rates. There is doubt that such a situation will help promote sustainable agricultural development. Smallholders in Kenya have expressed their dismay at the credit situation they are facing, insisting that it was the main cause of low agriculture productivity.There is limited data available on lending from microfinance to small scale agriculture for in sub-Saharan Africa. It is,however, estimated that less than 10 per cent of total lending by commercial financial institutions in sub-Saharan Africa goes to agriculture with large scale farmers as core benefactors (Mhlanga, 2010). Figure 1 provides an overview of the lending pattern of commercial banks to agriculture in a number of selected sub-Saharan African countries between 1995 and 2008. The graph shows that lending in a number of selected sub-Saharan countries has experienced modest growth with some cases of slight decline.

Figure 1: Value of commercial Bank lending to the agricultural sector of selected countries in Africa 1995 -2008 (USD million)

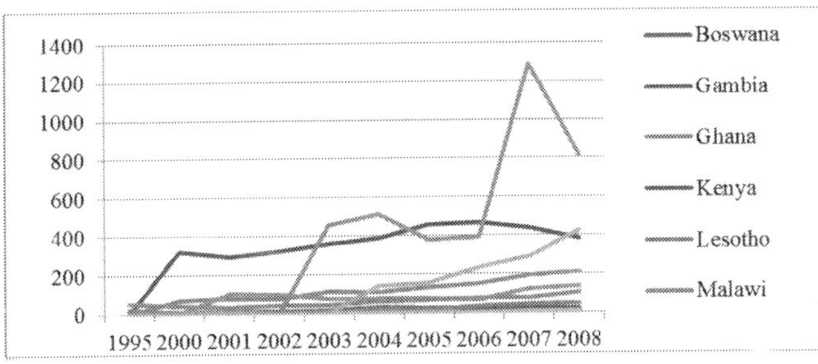

Source: Authors calculation using the data of Mhlanga (2010)

Although microfinance institutions (MFIs) have helped facilitate credit to the rural smallholders by adopting a different approach based on business's cash flow evaluation or income as eligible collateral substitute, its impact has not been widespread (Salami et al., 2010). Co-operatives in rural areas have also helped spread agricultural credit. This is because information complied on respective members by the organization is useful not only in the loan assessment process but also repayment due to peer pressure.Furthermore, financial institutions do not have to engage in high infrastructure costs like in the case of institutional set-up (Admasuand Paul, 2010). Financial institutions can thus reduce transaction costs by aligning operations to those of the co-operatives.

Eighty three percent of enterprises and firms in sub Saharan Africa were obliged to provide a form of security or deposit before they were granted loans by financial institutions (The World Bank, 2006). One major reason deterring commercial and for-profit financial institutions from disbursing credit to sub-Saharan Africa (smallholders) or small-scale agriculture is the lack of proper collateral and their documentation (Beck *et al.*, 2009). Sub-Saharan African agricultural production is characterized by a disproportionately large fraction of agricultural output which is in the hands of smallholder farmers. The average land holding is about one to three hectares, thus agricultural asset (farmlands) is available in rural areas, a number of smallholders however lease or rent farmlands for cultivation (Ogunlela and Mukhtar, 2009). Collateral plays a major role in credit risk and capital requirement of an agriculture portfolio (Katchova and Barry, 2005). Credit risk variables such the probability of default (PD) and capital adequacy′s value-at-risk (VaR) have collateral evaluation as an integral element (Katchova and Barry, 2005). PD estimates the probability that an individual farmer will not be able to meet his/her obligation, in other words, the likelihood that the farmer′s asset will fall below the farmer′s debt. Value-at-risk (VaR) estimate probability distribution of credit losses conditional on portfolio composition (Thonabauer and Nosslinger, 2004). If the asset valuation of the rural smallholders at the initial period is perceived to be zero, default already occurs as at the time that the smallholder applied for the loan. This method of modelling credit risk does marginalize small-scale agribusiness and/or smallholders due to the high risk and even higher capital requirements that the commercial banks would face in the event of lack of collateral. However, the conventional method of modelling agriculture risk and lending may be discriminatory toward sustainable agriculture which offers stable income eligible as collateral apart from the conventional farmland.

The difficulty project investors face, especially in developing countries, in balancing profitability with the needs of people NGOs etc. can be tackled by increasing local participation, long-term interests of projects, fair and competitive long-term price and uniform conservative environmental standard as well as contributing to peace and stability (Wagner, 2004). Financing of projects which are developed to be socially and environmentally sustainable is a way financial institutions can and are contributing to sustainable agriculture and development. The Environmental Social Governance (ESG) factor which remains an indicator for sustainable investments, amongst others investment decision-making, may however not be quantitative but rather qualitative in nature (Bassen and Kovacs, 2008). Financial institutions in Nigeria and South Africa are working to improve their ESG in terms of lending and investment under the Equator Principles (a comprehensive set of environmental and social guidelines for the financing of projects over US$50 million) although progress has been slow (UNEP, 2007a). The joint venture between Diamond bank (Nigeria), the international financial corporation (IFC) and shell petroleum, were only ecosystem friendly efficient businesses profit from an investment fund of ca. US$1.5 million and Ned-bank's Green Trust (South Africa) with US$770,000 in fund which focuses mainly on agribusiness are examples of how financial institutions are improving their ESG (UNEP, 2004). The total socially responsible investment fund for South Africa is estimated to be ca. US$1.4 billion. Although microfinance investment vehicles (MIVs) are promoting the use of ESG, enough is not being done across sub-Saharan Africa (CGAP, 2010a). With MIV concentrated in Eastern Europe and Central Asia (43 per cent), allocated investment to Africa (6 per cent) is quite minimal presumably due to the risks involved (CGAP, 2010b).

4. Sustainable Agriculture Opportunities for Sub-Saharan Credit Institutions

Climate change due to the emission of greenhouse gases (GHGs) primarily carbon dioxide (CO_2) and other gases such as methane (CH_4) and nitrous oxide (N_2O) are a global threat. Agricultural is a contributor to GHGs, fertilizer usage, burning and felling burning of trees, land use change and animal dung releases CO_2 and N_2O. Activities such as afforestation and reforestation, the management of forest, soil, livestock, manure and land, sustainable biofuel production, energy efficiencies and biodiversity mitigates GHGs emissions (Seeberg-Elverfeldt, 2010). Sustainable agriculture practices provide a solution to climate change since it has the ability to store and capture CO_2. Certain rural smallholders and small-scale agribusinesses are therefore able to mitigation climate change by engaging in afforestation (reforestation). Carbon capture and storage is economically viable due to the ability to trade each ton of sequestrated carbon as an entity on the secondary market similar to those on the stock exchange. "*Taungya*" is an example of sustainable small-scale agroforestry practice; *Taungya* is the combined cultivation of timber trees and horticultural crops (Harrison, 2009). Others nitrogen fixing trees with energy potential such as *Calliandra* and *Sesbania sesban*, Jatropha curcus and Pongamia pinnata can be used in small-scale agroforestry with diverse benefits (Rossi and Lambrou, 2009). A variety of small-scale sustainable agroforestry by smallholders can be beneficial not only in the adaptation and mitigation of climate change, but could help improve food and energy security if well propagated (Seeberg-Elverfeldt, 2010).

There is need to explore not only the direct benefits small-scale agroforestry may provide to smallholders such as yield improvement and increased productivity, but also the indirect benefit such as capital formation and increase welfare which may have positive impact on banks loans or agroforestry projects decision-making. This paper elaborates on Jatropha agroforestry, whereby an estimated 120,000 hectares is being cultivated in Africa by both corporate and small-scale farmers, Asia and Latin America have 760,000 and 20,000 hectares respectively (Brittaine and Lutaladio, 2010). The Jatropha tree has a high level of toxicity and is also invasive nature.Its oil production varies between 400 and 2,200 litres per hectares (Sielhorst *et al.*, 2008). Although Jatropha seed has been in use in sub-Saharan Africa for ca. 50 years, predominantly in the making of soap and lighting, its socio-economic evaluation has being limited. Another aspect that has received less attention apart from the indigenous use is the carbon sequestrated through its cultivation in small-scale agroforestry (Hellings *at el.*, 2012). It is argued that promoting carbon sink either through the Kyoto protocol or voluntary market provides a promising avenue to address the north-south equity issues, while noting that the conservation practices is beneficial to smallholders depending on the practice and household endowment (Tschakert, 2004). It is important to point out that the study above also emphasizes the need to assist smallholders financially, technically etc. in their pursuit of sustainability. To illustrate capture and storage (environmental service), a 3 hectares small-scale agroforestry with Jatropha may range from between 1100 and 2500 trees depending on density (Benge, 2006). Small-scale cultivation of Jatropha on marginal land or wasteland compared to large scale cultivation, although characterized by non-optimal growth, may be socio-economically viable. The sequestration potential of the small-scale cultivation is estimated to be 5.5 t CO2 ha^{-1} $year^{-1}$ (Abhilash *et al.*, 2010). The carbon sequestrated is a mitigation measure which is eligible for compensation in the form of carbon credit. the possible annual revenue from trading the sequestration mention above in 2007 on the clean development mechanism (CDM = US$11 → CO_2e) and voluntary market (VCS = US$12.5 → CO_2e) is equivalent to US$60 and US$68.75 respectively while estimated cost of carbon market participation of a micro-scale *project* (< 5000 ton CO_2/yr) for potential investor is ca. US$65,000 for the CDM and ca. US$25,000 for the VCS which includes negotiation, project approval, project monitoring, verification, and insurance costs (Green Markets International, 2007; Lipper at el., 2011; De Pinto *at el.*, 2010). Jatropha medicated soap production due to the simple local production techniques is observed in sub-Saharan Africa to be highly profitable and provides additional revenue to low-income earners (Eijck, 2007). The amount of soap derived from 13 litres of Jatropha oil, given an estimated five working hours, is 4.7k g (Henning, 2004). Seed cake residue which is a by-product of soap production serves as an organic fertilizer, applying this type of fertilizer on farmland is a way to generate carbon credit because it does not contribute to the release of N_2O (Brittaine and Lutaladio, 2010).

Model A

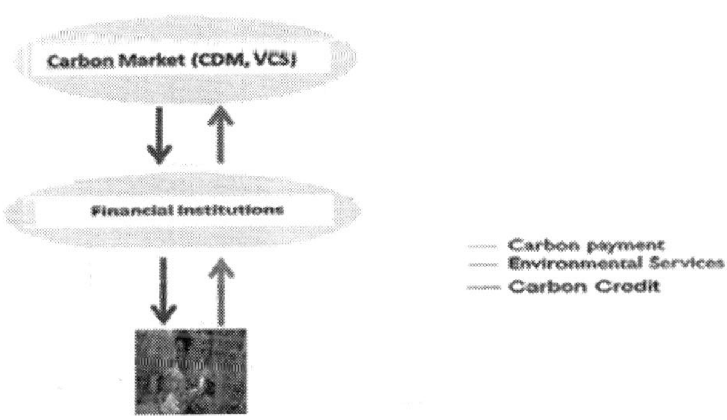

Source: Author

Carbon trading presents a win-win situation for both carbon buyers and sellers especially for sustainable smallholders in developing countries (The World Bank, 2011a). Some literatures out on how to create a broker between the demand and the supply of carbon certificates via emission trading championed by financial institutions are reviewed and models of climate-agricultural based investments and lending are introduce hereafter."Carbon trading partners are to be linked by an investment vehicle with adequate financial capital willing to dedicate substantial capital to the project" (Perez *et al.*, 2007). "Accumulating and storing carbon may also yield higher returns whereby these returns may come with a lag or delay, if this is the case farmers may require a positive financial incentive, such as a loan, to be encouraged to bear the fixed and variable costs of adopting and maintaining conservation practices" (Antle and Diagana, 2003). The argument of the authors is that the carbon market can act as the financier of sustainability if stable financial institutions are in place.

Microfinance institutions can reduce credit risk and promote sustainability improving on their ESG factor if there are able to target rural climate certified smallholders. The voluntary certification process for smallholders, although still in its infancy, is starting to gain recognition as conservation practices have not only resulted in stable income through improved yield but also carbon revenues (The World Bank, 2011b). The database of verified carbon standards puts the number of agriculture, forestry and land use related project observed in countries such as Kenya, Tanzania, Uganda, South Africa, Mali and Madagascar with active participation from NGOs at 11 (VCS, 2012). Since the support for the inclusion of social and environmental risk in credit risk analysis of African banks is on the increase, smallholder carbon credit inclusion in risk calculation may lead to more lending (UNEP, 2007b). Carbon credit on one hand may to some extent improve smallholders asset base making them less riskier as far as calculation of the probability of default is concern and on the other hand may help free up capital which would otherwise be tied to capital adequacy.

Model B

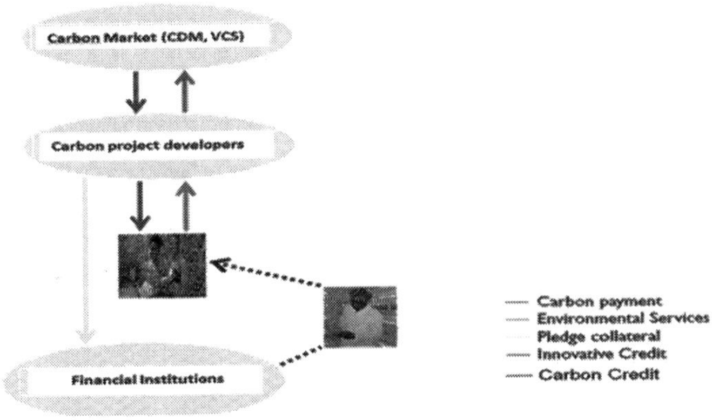

Source: Author

5. Policy Implication

The policies on (small-scale) agriculture drafted by Africa governments and international organization and donors which seeks the development of agricultural research, technology dissemination and adoption to sustain long-term productivity growth and national poverty reduction have not adequately addressed the agriculture and rural development (UKFG, 2012). Theoretically, where there is efficiency in small scale agriculture, government intervention in agriculture should be minimal rather market forces should prevail, thus the focus of policy interventions should be to offer an enabling economic environment for market-led development (Daio, 2007). The author further argues that the existence of institutional and market failures, ultimately leading to poverty trap, would however not permit such hands–off approach. Diversifying the income of poor smallholders in developing countries are slim, effective policies which may help broaden this opportunity have to be formulated. This formulation should give rise to innovative institutional arrangements between the public and private sector which guarantees the growth of both sectors.

The existence of a equator principle for financial institution to foster development based on the framework of international finance corporation performance standards on social and environmental sustainability and on the world bank group environmental, health, and safety guidelines should pave way for a sub-Saharan Africa version directed at microfinance institutions (EP, 2011). The limited number of financial institutions from sub-Saharan Africa, namely six, who are signatory to the equator principle mainly from South Africa and Nigeria signals a need by the governments and national regulators to push for a bottom-up approach.

6. Conclusion

The problems confronting sub-Saharan agriculture and a move towards a sustainable agriculture are complex and far reaching. As the demand for and the prices of food and fossil fuel continue to surge, with Africa being the most vulnerable, concerns about sustainability and climate change are genuine. In order to ensure sustainable food and energy security, agriculture development in a sustainable manner has to be the main focus. With the food versus fuel and poverty elevation debate heating up in sun-Saharan Africa, a possible solution amongst others is in sustainable small-scale agroforestry with energy potential. The guidelines of international climate consensus such as the CDM and voluntary carbon markets are continuously amended to accommodate rural sustainable smallholders in developing countries. The eligibility for carbon credit certificate due to sustainable farming and resulting trading on carbon market means smallholders are recipient of stable long-term income. The active participation of financial and non-bank financial institutions on the carbon projects may provide climate financing capacity building and a base for the asset valuation of these local smallholders in credit risk management process. The opportunities that climate mitigation instruments presents can be used to reverse the trend of credit constraints due to lack of collateral (farm asset or stable income). A number of issues however need further investigation, such as the actual cost and revenue as well as risks of climate financing and sustainable lending through accurate case studies.

References

Abhilash, P. C., Srivastava, P., Jamil, S. and Singh, N. (2010), 'Revisited Jatropha curcas as an oil plant of multiple benefits: critical research needs and prospects for the future,' in *Environmental Science and Pollution Research*18:127–131

Admasu, A. and I. Paul (2010), 'Assessment on the mechanism and challenges of small scale agricultural credit from commercial banks in Ethiopia: the case of Adaá liben woreda Ethiopia,' in *Journal of Sustainable Development in Africa,* 12(3): 323 – 304.

Ahmed, A. (2005), 'The impact of finance and funding on technology adoption,' in *Journal of African Development,* 7 (1): 20-41.

Ahma, W. (2001) '*Revisiting the Regulatory and Supervision Framework of the Micro-finance Industry in Ethiopia;*' Norway: Drylands Coordination Group / Noragric.

Antle, J. M. and B. Diagana. (2003), 'Creating Incentives for the adoption of sustainable agriculture practices in developing countries: The role of soil carbon sequestration,' in *American Journal of Agriculturul Economics,* 85(5): 1178–1184.

Arun, T. and Murinde, V. (2010), 'Microfinance regulation and social protection,' *Paper presented at the Conference Experiences and lessons from social protection programmes across the developing world: what role for the EU,* 17-18 June, Paris, France.

Augsburg, B., Schmidt, J.P. and Krishnaswamy K. (2011), 'Free and open source software for microfinance: increasing efficiency and extending benefits to the poor,' In Ashta, A. (ed.),*Advanced Technologies for Microfinance: Solutions and* Challenges, Hershey, PA:IGI Global.

Balkenhol, B. (1991), 'Savings, Credit, and the Poor: What Has the ILO to Do with the Financial Sector,' in *International Labour Review,* 130(5-6): 645-655.

Bassen, A., and Kovacs, A., (2008), 'Environmental, social and governance key performance indicators from a capital market perspective,' in *The Journal for Business, Economics and Ethics,* 9(2): 182-192.

Basu, A., Blavyand, R. and Yulek, M. (2004), 'Microfinance in Africa: Experience and Lessons from Selected African Countries,' in *International Monetary Fund (IMF), Working Paper,* 174: 1 – 24.

Benge, M. (2006), 'Assessment of the potential of Jatropha curcas, (biodiesel tree) for energy production and other uses in developing countries;' *Ascension Publishing,* 20 August 2001. Available at: http://www.ascension-publishing.com/BIZ/jatropha.pdf.

Beck, T., Fuchs, M. and Uy, M. (2009), 'Finance in Africa: Achievements and Challenges,' in *The World Bank Policy Research Working Paper,* 5020: 31 – 47.

BIS (2010), *Microfinance activities and core principle for effective banking supervision,* Basel: Bank for international settlement (BIS) communication.

Branca, G., Tennigkeit, T., Mann, W., and Lipper, L. (2011), *Identifying Opportunities for Climate-Smart Agriculture Investments in Africa,* Rome: Food and Agriculture Organization of the United Nation and World Bank.

Brittaine, R., and Lutaladio N. (2010), *Jatropha: A Smallholder Bioenergy Crop: The Potential for Pro-Poor Development,* Rome: United Nations.

Brownbridge, M. (1998), 'The Causes of Financial Distress in Local Banks in Africa and implication for prudential Policy,' in *United Nations Conference on Trade and Development,* 132: 1– 40.

Campion, A. and White, V. (2001), 'NGO Transformation,' *microfinancegateway,* June 9. Available at:http://www.microfinancegateway.org/gm/document-1.9.28745/19230_N_073.pdf.

Charitonenko, S. and Campion, A. (2003), 'Expanding Commercial Micro Finance in Rural Areas: Constraints and Opportunities;' *Chemonics,* June 9. Available at:www.basis.wisc.edu/live/rfc/cs_05a.pdf

CGAP (2010a), 'CGAP Awards Recognize Three Microfinance Investors for Social Commitment,' *CGAP,* June 9. Available at:http://www.cgap.org/p/site/c/template.rc/1.26.14337/.

CGAP (2010b), MIV Survey Report Market Data and Peer Group Analysis, *CGAP*, June 9. Available at: http://www.cgap.org/gm/document-1.9.47373/CGAP_2010_MIV_Survey_Report.pdf.

Christiaensen, L.J. and Demery L. (2007), *'Down to Earth: Agriculture and Poverty Reduction in Africa,'* Washington D.C.: The World Bank.

Comninos, A., Esselaar, S., Ndiwalana, A. and Stork, C. (2008), 'M-banking the Unbanked,'in*Research ICT Africa* 1(4): 1 – 20.

Conway, J. and Stevens, A. (2012), 'Industry Survey of Financial Advisors on Sustainable and Impact Investing' gatewayto-impact.org, June 9. Available at: http://www.gatewaystoimpact.org/images/ gatewaystoimpact.pdf.

Cull, R., Demirgüç-Kunt, A. and Morduch, J. (2008), *Microfinance Meets the Market,* New York: The World bank.

De Pinto, A., Magalhaes, M. and Ringler, C. (2010), 'Potential of Carbon Markets for Small Farmers: A Literature Review,' in *IFPRI Discussion Paper*, 1004:1–36.

Diagne, A., Simtowe, F. and Zelle, M. (2009), 'The impact of credit constraints on the adoption of hybrid maize in Malawi,' in *Review of Agricultural and Environmental Studies - Revue d'Etudes en Agriculture et Environnement*, 90(1): 5-22.

Eijck, V. J. (2007), *Transition towards Jatropha biofuels in Tanzania? An analysis with Strategic Niche Management,'* Leiden: African Studies Centre.

EP (2011), 'Members and Reporting, Equator Principle,' June 9. Available at:http://www.equator-principles.com/index.php/members-reporting.

Feder, G. and Umali, D.L. (1993), 'The Adoption of Agricultural Innovations: A Review,' in *Technological Forecasting and Social Change*, (43): 215-239.

Fernandez-Cornejo, J. and McBride, W.D. (2002), 'Adoption of Bioengineered Crops,' in *agriculture economic report No. 810*, Washington, D.C: U.S. department of agriculture.

Fernando, N.A. (2008), 'Managing Microfinance Risks: Some Observations and Suggestions,' in *Asian Journal of Agriculture and Development*, 4(2): 1-22.

Flamini, V., McDonald, C. and Schumacher L. (2009), 'The Determinants of Commercial Bank Profitability in Sub-Saharan Africa,' in *IMF Working Paper*, 09(15): 1-32.

Furrer, B., Hoffmann, V. and Swoboda, M. (2009), 'Banking and Climate Change:Opportunities and Risks An Analysis of Climate Strategies in more than 100 banks worldwide', *ETH Zurich and SAM*, June 9. Available at: http://ec.europa.eu/enterprise/policies/sustainable-business/corporate-social-responsibility/reporting-disclosure/swedish-presidency/files/surveys_and_reports/banking_and_climate_ change-_sam_group_en.pdf.

Gonzalez-Vega, C. (1993), 'From Policies, to Technologies, to Organizations: The Evolution of the Ohio State University Vision of Rural Financial Markets,' *Economics and Sociology Occasional Paper No. 2062*, Columbus, Ohio: The Ohio State University.

Gonzalez-Vega, C. and Graham, D.H. (1995), 'State – Owned agricultural Development Banks: Lessons and Opportunities,' *Economics and Sociology Occasional Paper No. 2245*, Columbus, Ohio: The Ohio State University.

Green Markets International (2007), 'The Voluntary Carbon Market: Status and Potential to Advance Sustainable Energy Activities', *Greenmarkets*,June 9. Available at: http://www.green-markets.org/Downloads/vCarbon.pdf.

Harrison, S.R., Jesusco, J., Sevare, M. and Mangaoang, E. (2009), *Agroforestry farming practices of smallholders in Leyte and implications for agroforestry systems design*, Queensland: The University of Queensland.

Hellings, B.F., Romijn, H.A. and Franken, Y.J. (2012), 'Carbon storage in Jatropha curcas trees in Northern Tanzania,' in-*Fuels from Agriculture in Communal Technology (FACT)*, (1): 1–25.

Henning, R.K. (2004), 'Jatropha curcas in Africa: Assessment of the impact of the dissemination of "the Jatropha System" on the ecology of the rural area and the social and economic situation of the rural population (target group) in selected countries in Africa,' *GFU*, June 9. Available at: http://www.underutilized-species.org/documents/publications/jatropha_curcas_africa.pdf.

Hishigsuren, G. (2006), 'Information and Communication Technology and Microfinance: Options for Mongolia,' in *ADB Institute Discussion Paper*, (42):1–33.

Jones, W.A. and Hiller, T.B. (2010), 'Achieving Scale in Investment Global Risk and Opportunity from Climate Change', *UK's First Climate Change Risk Assessment (CCRA)*,June 9. Available at: http://www.foresight.gov.uk/index.asp.

Katchova, A. and Barry, P. (2005), 'Credit Risk Models and Agricultural Lending,' in *American Journal of Agricultural Economics*, 87(1): 194-205.

Kiyota, H. (2009), 'Efficiency of Commercial Banks in Sub-Saharan Africa: A Comparative Analysis of Domestic and Foreign Banks,' *Paper presented at the CSAE Conference*, 22 - 24 March, University of Oxford, UK.

Lafourcade, A., Isern, J., Mwangi, P. and Brown M. (2005), 'Overview of the Outreach and Financial Performance of Microfinance Institutions in Africa', *Microfinance Information eXchange (MIX)*, Nov. 10. Available at: http://www.ruralfinancenetwork.org/pubs/MIX%20Africa_Data_Study.pdf.

Lipper, L., Neves, B., Wilkes, A., Tennigkeit, T., Gerber, P., Henderson, B., Branca, G. and Mann, W. (2011), *Climate Change Mitigation Finance for Smallholder Agriculture: A guide book to harvesting soil carbon sequestration benefits*, Rome: Food and Agriculture Organization of the United Nations.

Mhlanga, N. (2010), *'Private sector agribusiness investment in sub-Saharan,'* Rome: Food and Agricultural Organization of the United Nations.

MIX (2008), 'Sub-Saharan Africa Microfinance Analysis and Benchmarking Report,' *Microfinance Information eXchange (MIX)*,' June 9. Available at: http://www.themix.org/sites/default/files/2008%20Africa%20Microfinance%20Analysis%20and%20Benchmarking%20Report%20-%20English.pdf.

MIX (2010), 'Sub-Saharan Africa Microfinance Analysis and Benchmarking Report,' *Microfinance Information eXchange (MIX),'* June 9. Available at: http://www.themix.org/sites/default/files/096_MIX_Africa%20Report_05-05-2011.pdf.

Ogunlela, Y. and Mukhtar, A. (2009), 'Gender Issues in Agriculture and Rural Development in Nigeria: The Role of Women,' in *Humanity and Social Sciences Journal*, 4(1): 19-30.

Perez, C., Roncoli, C., Neely C. and Steiner J.L. (2007), 'Can carbon sequestration markets benefit low-income producers in semi-arid Africa? Potentials and challenges,' in *Agricultural Systems*, (94): 2-12.

Porteous, D. (2006), *The Enabling Environment for Mobile Banking in Africa*,Boston: Department for international development.

Randhawa, B. and Gallardo, J. (2003), 'Microfinance Regulation in Tanzania: Implications for Development and Performance of the Industry', in *Africa Region Working Paper Series*, (51): 1-48

Rossi, A. and Lambrou, Y. (2009), *Making sustainable biofuel work for smallholder farmers and rural households: issues and perspectives*, Rome: FAO.

Salami, A. Kamara, B. and Brixiova Z. (2010) *'Smallholder Agriculture in East Africa: Trends, Constraints and Opportunities,'* Tunis: African development bank (ADB).

Schneider, S.H., S. Semenov, A. Patwardhan, I. Burton, C.H.D. Magadza, M. Oppenheimer, A.B. Pittock, A. Rahman, J.B. Smith, A. Suarez and F. Yamin, (2007), 'Assessing key vulnerabilities and the risk from climate change', In: *Climate Change 2007: Impacts, Adaptation and Vulnerability*, Contribution of Working Group II to the Fourth Assessment Report of the Intergovernmental Panel on Climate Change, M.L. Parry, O.F. Canziani, J.P. Palutikof, P.J. van der Linden and C.E. Hanson (eds.), Cambridge University Press, Cambridge, UK, 779-810.

Seeberg-Elverfeldt, C. (2010),'Carbon finance possibilities for Agriculture, Forestry and Other Land Use Projects in a Smallholder Context,' in *Environmental and natural resources management working paper 34*. Rome: United Nations.

Sielhorst, S., Molenaar J. W. and Offermans D. (2008), 'Biofuels in Africa An assessment of risks and benefits for African wetlands,' *Wetlands international,* June 9. Available at: http://www.wetlands.org/LinkClick.aspx?fileticket=vPIIvbwvqTs%3D&tabid=56.

Steel, W.F. and Andah, D.O. (2003), 'Rural and microfinance regulation in Ghana: Implications for development and performance of the industry,' *World Bank Africa Regional Working paper Series No. 49,* Washington, D.C.: The World Bank.

The World Bank (2006), *Africa Region: Making Finance Work for Africa,* Washington D.C.: The World Bank.

The World Bank (2007), *World Bank Assistance to Agriculture in Sub-Saharan Africa: an IEG review,* Washington, D.C.: The World Bank.

The World Bank (2011a), 'First African Emission Reductions Purchase Agreement For Soil Carbon Signed in The Hague' *Press Release No: 2011 /165/SDN,* June 9. Available at: http://web.worldbank.org/WBSITE/EXTERNAL/NEWS/0contentMDK:22753334~pagePK:34370~piPK:34424~theSitePK:4607,00.html.

The World Bank (2011b), 'Triple Win of Climate-Smart Agriculture put into Practice,' June 9. Available at: http://web.worldbank.org/WBSITE/EXTERNAL/TOPICS/EXTSDNET/0contentMDK:22842518~menuPK:64885113~pagePK:7278667~piPK:64911824~theSitePK:5929282,00.html.

Thonabauer, G. and Nosslinger B. (2004), *Rating Models and Validation: Guidelines on Credit Risk Management,* Vienna: Oesterreichische Nationalbank.

Tschakert, P. (2004), 'Carbon for farmers: Assessing the potential for soil carbon sequestration in the old peanut basin of Senegal,' in *Climatic Change,* 67(2-3): 273–290.

UCB (2010), 'Uganda's Experience in Regulating Microfinance Deposit-taking Institutions,' *Paper presented at the Microfinance Regulation – Who Benefits? Conference,* 15 - 17 March, Dhaka, Bangladesh.

UNEP (2004), *Sustainability Banking in Africa,* Nairobi: The United Nations Environment Program Finance Initiative.

UNEP (2007a), 'Africa's Banks Starting to Rise to Challenge of Wider Environmental, Social and Governance Concerns,' *UNEP News Release 2007/22,* June 9. Available at: http://www.unep.org/Documents.Multilingual/Default.Print.asp?DocumentID=512&ArticleID=5613&l=en.

UNEP (2007b), *Banking on Value: A New Approach to Credit Risk in Africa,* South Africa: United Nations Environment Program Finance Initiative (UNEP FI) and African Task Force (ATF).

UKFG (2012), 'More Aid for African Agriculture Policy implications for small-scale farmers,' *UK Food Group,* June 9. Available at: http://www.ukfg.org.uk/docs/More_Aid_for_African_Agriculture_MAIN_REPORT.pdf.

VCS (2012), 'The VCS Project Database,' *Voluntary carbon standard,* June 9, Available at: http://www.vcsprojectdatabase.org/.

Verhoef, G. (2005), 'Globalization, concentration and regulation: Trends in the globalization of South African banking, 1990-2005,' *University of Johannesburg,* June 9. Available at: http://ehsanz.econ.usyd.edu.au/papers/Verhoef.pdf.

Wagner, D. (2004), *Investing in Stability: Conflict Risk, Environmental Challenges, and the Bottom Line,* United Nations Environment Program and International Institute for Sustainable Development.

Wisniwski, S. (1998), 'Savings in the Context of Microfinance – Lessons Learned from Six Deposit-Taking Institutions,' *Paper presented at the Interamerican Microenterprise forum,* 26-28 March Mexico-City, Mexico.

WSBI (2008) Microfinance services by savings banks in Africa: The sleeping giants have started moving, but where are they going? *World savings banks and institute,* June 9. Available at: http://www.wsbi.org/uploadedFiles/Publications_and_Research_%28WSBI_only%29/microfinance%20leadership%20of%20WSBI%20African%20members%20GB%20screenview5.pdf.

Chapter 13

Post Scriptum : Researching Innovation Systems at the Sectoral Level

Mammo Muchie & Angathevar Baskaran

Varieties in the presentation of systems of innovation perspectives continue. In this book the contributions were mainly innovation dynamics and systems at the sectoral level. We are now at the time when development economics is seen to have not succeeded to provide the intellectual lens to remove poverty, unemployment and inequality. Now we have also the climate change risk that the world Governements have signed SDGs to realise a future by 2030 that will be inclusive, humane with social and environmental wellbeing.

We recognise development economics evolved in the post-war world when many colonies were granted political independence and the issue of development confronted both the elites and the former colonial powers. It continued during the Cold War and still exists to this day. In the mean time, the approach of systems of innovation is now over a generation old. The national system of innovation, the most popularized version of the innovation system framework originated in the North. It evolved in search of how to organize science, technology, engineering, mathematics and innovation systems effectively for both policy learning and managing economic development. Since 1980s theories on innovation and their use have gradually expanded their focus and complexity. From the initial focus on the individual firm or entrepreneur they expanded to include the locality, community, city, region, industry, sector, nation and the globe. They focused on incentives, intellectual property rights and on the national system of regulations, institutions, human capital and government policies and programmes. The focus also included regional level, sectoral or local level systems of innovations. In other words, initial perception that innovation is basically an individual act of learning by a firm or entrepreneur has expanded to include the larger system (consisting of various institutions, policy framework, incentives etc.) in which this act occurs. It is now widely viewed and accepted that innovation is a process, which is not linear as it involves continuous interactivity between various actors and factors at various levels from local and sectoral to regional, national and global levels. The use of the concept of systems of innovation has grown and proliferated over the years. In the area of spatial we can identify global innovation systems, national innovation systems, regional and sub-regional innovation systems, sectoral, local and city innovation systems. In the area of industry and technology specific innovation systems, we can see research and studies focused on specific industrial sectors such as manufacturing, telecommunications, automotive, agro-food and service and specific technology focused such as biotechnology, information and communication technology (ICT), and electronics. In terms of innovations types we can see the focus of studies on product innovation, process innovation, service innovation and organisational innovation and so on. Similarly, research that are focused on levels of technological or innovation complexities involving incremetal process modifications, modular, architectural, revolutionary, radical, systemic and paradigm shift.

We can also broadly see studies focusing on innovations driven by social objectives or economic motivations or objectives. The system of innovation is a concept utilized to describe the relationship between internal processes in firms and external processes in the wider environment in the context of knowledge creation, diffusion, and transfer. In the developed world's economic development the innovation system framework is based on empirical evidence on technologies, knowledge, innovation and learning. In many developing economies we cannot expect the System of Innovation to be copied. It has to be made by taking the different contexts into consideration. There is a need to re-frame innovation system to make it relevant to embrace the development economics interactions in a specific and empirically valid way amongst relevant stakeholders that bring knowledge, learning, innovation and competence building as critical drivers to promote developmental-transformational outcomes. There is a need to identify how given development economics how the innovation system concept features can be re-worked. To begin with this is how innovation system has been conceptualized. There are many types of interactions that take place both within firms and outside firms.

Among the various elements the concept of innovation system identifies is related to variation of the elements that constitute parts of a system. A system of innovation is a concept to distinguish the most significant interactions from less significant ones in relation to actors, activities, and institutions in the process of knowledge creation, exchange, diffusion and transfer. Interactions that promote or hinder knowledge diffusion, exchange or transfer are even more relevant in the development economics of innovation system constructions. The space could vary (local or national, or regional or global). The sector/industry could vary. The concept of innovation relates knowledge creation, diffusion and transfer to the actors, activities, institutions, spaces and their interactions. The interactions in the process of acquisition, diffusion and transfer of knowledge can form different degrees and levels of systemic properties and functions around the creation of innovation as the core. The concept of innovation system captures the specific interaction of 'innovation-knowledge' as the independent variable and spatial, sectoral and institutional arrangements as the dependent variable in the context of the activities and actions to bring about transformation and development.

As a critique, the concept of 'system of innovation' can be

used to challenge ideas about wealth accumulation based exclusively on static comparative advantage without products, markets, organizations, processes, innovation and learning imparting development features to a given national economy. Many developing countries depend on one or two main commodities for export, and they are advised to specialize in these commodities to earn the foreign exchange that may be ploughed back into the economy.

Development is conceptualized as a consequence of the incremental growth that this export-orientated strategy based on the comparative advantage of agricultural and mineral produces might yield. The system of innovation perspective questions the significance of the development features, development dynamics and development effects of such an export-promoting specialization development strategy. It points direction and policy to the significance of the co-evolution of technologies, learning and institutions and incentives by bringing about systemic and significant interactions of the social, economic and political domains in order to imbed development features and development effects by diversifying the product and process base of a national economy. As a metaphor 'system of innovation' orientates actors to integrate knowledge, innovation and learning to solve problems based on their own resources with self-reliance rather than resorting to dependency.

Innovation system suggests that ideas, the domestic actors, institutions and incentives must interact in order to create new opportunities in production, distribution, markets and circulation. This can inculcate a mental attitude of 'can do it yourself', rather than waiting for external impulses to create dynamic activities in the economy. A key attribute of the innovation system concept is the focus it provides in framing problems and the value it signifies to the domain of reality that mainstream economics neglects or underemphasizes. The innovation system concept makes central, institutions, histories, territories, technologies, organizations and nations that are often neglected and treated as a residual in mainstream neo-classical economics. The concept has evolved by putting innovation and learning at the heart of the economics of development.

In order to address these challenges that development economic has been trying to address for a half-century, we propose that what is needed is not merely Innovation System (IS) conceptual framework, but a more relevant and immediate focusing framework of ISDS (Innovation for Sustainable Development System.) We started to develop research that has led us to promote the ISDS framework by exploring how the innovation and learning approach captured by Innovation System can be applicable for economies in the developing world in general including Africa in particular.

Development is driven by the intellectual achievements, discoveries, inventions, transformations and progress that a nation has accumulated in its history. The innovation system framework suggests that theoretically development is not only a process of production, but also it is the process of innovation. In addition developing economies are not mere users of technology, but also their development should be conceptualized on the domain of active generators or producers of technology. Developing economies are not therefore passive recipients of technology from the industrialized economies. The assumption that development can be engineered or steered by technologies that can be transferred from those who control knowledge through intellectual property is a seriously inadequate proposition.

There is also the problem that technology transfer cannot be assumed to take place easily. Often the proprietors of technology control knowledge, they do no readily spread it. This provides an additional reason for organizing a national system of innovation. Thus there is no alternative that those who wish to use technology must be prepared also to learn to produce it. That is what africa should do : become the science, technology, engineering, mathematics, knowledge, talent, invention, innovation and entrepreneurship hub of the world. Africa must become the destination not the departure zone of invention, creativity, innovation and knowledge. A necessary condition for development from an innovation system perspective is that those who wish to embark on a sustainable developmental trajectory cannot afford to shy away from trying to develop by being producers of technology, and not stay merely as diffusers, users, absorbers and implementers of technologies produced elsewhere.

Finally sectoral systems of innovation should address how technology and innovation evolve at various levels, not only at the sectoral level. The key is to comprehend how at the sectoral level the contribution of how the sustainable development economics of innovation system makes development a pillar where specific sectors become or specialize as both producers and absorbers, creators and diffusers of innovation. In many developing econmies, it is likely that there may be more diffusers and absorbers than creators and producers of technology at present time. Of course both producers and users of technology undertake different types of innovation. The broad classification of economies as those who are the technology producers as the innovators, those that are absorbers of created technologies elsewhere as the diffusers, and those that may be excluded from being either innovators or diffusers create how a new sustainable development economics of innovation system may capture the dynamics and variation of the development process. The contributions in this book highlight these issues encouraging more research and knowledge to be produced.

Notes on Contributors

Mammo Muchie

DST/NRF SARChI Professor (Innovation & Development), Tshwane University of Technology, South Africa; Technology and Management Centre for Development (TMCD), Oxford University, UK; Associate Faculty Professor, Sussex University, UK; Adama Science and Technology University (ASTU), Ethiopia. Email: MuchieM@tut.ac.za

Angathevar Baskaran

Associate Professor, Department of Development Studies, Faculty of Economics and Administration, University of Malaya, Malaysia; Senior Research Associate of SARChI (Innovation & Development), Tshwane University of Technology, Pretoria, South Africa. Email: baskaran@um.edu.my & anga_bas@yahoo.co.uk

John Olatunji Adeoti

Economic Policy Research Department, Nigerian Institute of Social and Economic Research (NISER), Oyo Road, Ojoo, P.M.B. 5 UIPO, Ibadan, Nigeria. Email: Adeotij@yahoo.com

Adetola Ibidunni Adeoti

Department of Agricultural Economics, University of Ibadan, Ibadan, Nigeria. Email: Jadeoti@yahoo.co.uk

Niringiye Aggrey

Lecturer, School of Economics, College of Business and Management, Makerere University, Kampala, Uganda. Emails:aggrey1970@yahoo.com & aniringiye@fema.mak.ac.ug

Sebastian Derwisch

System Dynamics Group, Department of Geography, University of Bergen, Postbox 7800, 5020 Bergen, Norway. Email: s.derwisch@cgiar.org

Birgit Kopainsky

System Dynamics Group, Department of Geography, University of Bergen, Postbox 7800, 5020 Bergen, Norway. Email: Birgit.kopainsky@flurygiuliani.ch

Siyanbola Willie O.,

National Centre for Technology Management, P.M.B. 012, Obafemi Awolowo University, Ile-Ife. Nigeria. Email:dg.ceo@nacetem.org

Olamade Olumuyiwa O.

National Centre for Technology Management, Nigeria.Email: muyiwaolamade@yahoo.com

Isola Oluseyi O.

National Centre for Technology Management, Nigeria.Email: seyiatzion@yahoo.com, seyiatzion@gmail.com

Egbetokun Abiodun A.

National Centre for Technology Management, Nigeria.Email: abiodun.egbetokun@nacetem.org

Aminu Bayawa Muhammad

Department of Pure & Applied Chemistry, Usmanu Danfodiyo University, Sokoto, Nigeria.Email: ambayawa@udusk.edu.ng; ambayawa@yahoo.co.uk

Abdullahi Abdu Zuru

Department of Pure & Applied Chemistry, Usmanu Danfodiyo University, Sokoto, Nigeria; Email:zurupeni@yahoo.com

Muhammad Gidado Liman

Department of Pure & Applied Chemistry, Usmanu Danfodiyo University, Sokoto, Nigeria; Email:mgliman@yahoo.com

Ahmad Ahmad Tijjani

Department of Pure & Applied Chemistry, Usmanu Danfodiyo University, Sokoto, Nigeria; Email:mawashiahmed@gmail.com

Uyilawa Usuanlele

Department of History, State University of New York, Oswego, NY.13126, Email: uyilawa.usuanlele@oswego.edu.

Lawrence K. Tshivhase

Entrepreneurial Studies, Tshwane University of Technology Business School, Pretoria, South Africa, Email:lawrence@degsa.co.za.

Zeleke Worku

Professor, Tshwane University of Technology Business School, Pretoria, South Africa, Email: WorkuZ@tut.ac.za.

Elias Madzudzo

Social Scientist, WorldFish Center, P.O. Box 51289, Ridgeway, Lusaka, Zambia, Email: e.madzudzo@cgiar.org or emadzudzo@gmail.com.

Oluseye O. Jegede

National Centre for Technology Management, Obafemi Awolowo University, Ile-Ife, Nigeria. Email: jegede.seye@yahoo.com

Abiodun A. Egbetokun

Post-Doc of DST/NRF Research Chair National Centre for Technology Management, Obafemi Awolowo University, Ile-Ife, Nigeria. E-mail: egbetokunaaAgmail.com

Willie O. Siyanbola

National Centre for Technology Management, Obafemi Awolowo University, Ile-Ife, Nigeria, E-mail: williesiyanbola@yahoo.com

Benjamin Olatunbosun

Bergische University of Wuppertal (BUW), Technical University of Munich (TUM), Germany. Email: emmanuel.benjamin@tum.de

Index

A

D

O

P

T

U

V

W